IRON IN THE PINES

IRON IN THE PINES

The Story of New Jersey's
Ghost Towns and Bog Iron

ARTHUR D. PIERCE

RUTGERS UNIVERSITY PRESS

New Brunswick, New Jersey

To

Margaret Sherwood Pierce

My Wife

With Love and Devotion

Library of Congress Cataloging in Publication Data

Pierce, Arthur D. (Arthur Dudley)
    Iron in the pines.

    Bibliography: p.

    Includes index.
        1. Cities and towns, Ruined, extinct, etc.—
New Jersey—Wharton State Forest.   2. Historic
sites—New Jersey—Wharton State Forest.   3. Iron
industry and trade—New Jersey—Wharton State
Forest—History.   4. Wharton State Forest (N.J.)—
History.   5. Iron industry and trade—New Jersey—
Wharton State Forest—History.   6. Batsto (N.J.)—
History.   I. Title.
F142.W47P54   1984   974.9   83-24739
ISBN 0-8135-0514-3

CONTENTS

Illustrations between pages 116 and 117

FOREWORD

Since his childhood New Jersey's "Wharton Estate" has been a happy hunting ground for the author, and his object in writing its story is to infect others with his enthusiasm for that part of the country. If, in addition, these chapters contribute in any way toward the State of New Jersey's efforts to conserve the natural resources of the area and to restore its historic communities, the years of research and exploration will be even more worthwhile.

The original spark for this book was generated by curiosity, a desire to know the truth about this remarkable region dominated by the "Wharton Estate." Its history has long been hazy. There have been an abundance of legend and a scarcity of fact. *Iron in the Pines* represents an effort to remedy that imbalance. While this book inevitably contains some material familiar to students of the Pine Barrens, much appears here in print for the first time. There is the last pathetic letter of Charles Read, New Jersey's exiled ironmaster; the real story of the "Kate Aylesford House"; solution of the "mystery" of the Wading River Forge and Slitting Mill; the actual location of the fabled Washington Tavern; the story of Atsion's two real estate booms and the paper mill which was built there; Batsto's long-shrouded origins, and fresh information on its colorful subsequent history. New also are much of the data on Martha Furnace, the extracts from the Batsto Store Books, and a good bit of the story of Pleasant Mills with its cotton and paper factories and the deeding of its historic church. While Pleasant Mills and Etna Furnace are not within the confines of the "Wharton Estate," and thus not State property, their history is part of the story of the area. Pleasant Mills is a twin village to Batsto, and Etna rounds out the chronicle of Charles Read's ill-starred empire.

Wherever possible, information has been obtained from primary sources. This has meant research in State and county records, often with only fragmentary leads to follow. It has involved checking fact after fact, statement after statement. Inevitably there remained gaps which could be filled only by relying upon the writings of others, most of which have been long out of print. The author would be the last to believe that no errors have crept into this book. He only hopes they have been few. But every effort has been made to distinguish among those facts which have been established, such information as remains in doubt, and that which is purely legendary. The author does not think the less of legends. They serve to light up fact with folklore. It is wise, however, to draw a line between the two, and it is doubly prudent to reject hearsay and gossip which bear no relation either to legend or to truth.

Historical research is a collective enterprise, and this book would have been impossible without the cooperation, encouragement, and labor of many people. Obvious is the debt the author owes to those mentioned in the text, especially to the late Charles S. Boyer and Dr. Carl R. Woodward, author of *Ploughs and Politicks*. Nathaniel Ewan made available the Martha Furnace Diary and countless notes and records, and helpfully read much of the manuscript. Mary and Raymond Baker provided the surviving records of the Pleasant Mills paper factory, much other valued assistance, and many warm and pleasant memories. Particular thanks are due for Raymond's generous gift of the design for the endpapers. Captain Charles Wilson, Chief of the Delaware River Port Authority Police and a veteran Pine Barrens explorer, applied his talents for investigation to historical research and generously offered the results of his labors to the author, who is particularly grateful for the Batsto Store Books and his copy of the Martha Furnace Diary. Another gold mine of information, the Wharton Estate records, was made available by the Girard Trust Corn Exchange Bank of Philadelphia, and a special debt is owed to George R. Thompson, who is in charge of real estate for the institution.

Gratitude is due also to Professor K. Braddock-Rogers of the University of Pennsylvania, for the reprints of articles describing his own research into the New Jersey iron industry and the Harris-

Foreword

ville-Martha area; to Howard R. Kemble, Trustee of the Camden County Historical Society, for help in many ways, especially in combing old newspaper files; to the Insurance Company of North America for the photograph of Joseph Ball and the information on Charles Pettit; to Dr. Henry H. Bisbee, of Burlington; Watson Buck, of Rancocas; Mrs. George Eckhardt, of Hammonton; Earle Haines, of Medford; Helen Henderson, of Medford Lakes; Mrs. John Herman, of Batsto; Harry Marvin, of the New Jersey State Highway Department; Dr. Richard P. McCormick, of Rutgers; Harvey Moore, of Riverton; Elliott Porter, of Green Bank; Susan Richards, of Bryn Mawr; the late Mrs. Harriet R. J. Robeson, of Atlanta, Georgia; Leeson Small, manager for the Wharton Estate; Mr. and Mrs. John Stewart, of Nesco; Leon E. Todd, of Medford Lakes; Mrs. Natalie Richards Townsend, of Haverford; and Joseph J. Truncer, administrative chief of the Wharton Survey for the State of New Jersey.

Courteous cooperation has been given by the staffs of the Atlantic, Burlington, and Camden County Historical Societies; the Historical Society of Pennsylvania; the Burlington County Library; the Recorder of Deeds offices in Atlantic, Burlington, and Gloucester Counties; and the New Jersey Secretary of State's office generally and Mr. Thomas Atkins particularly. The greatest contribution of all has been that of the author's wife, Margaret, who joined him in tramping the woods, measuring cellar holes, locating old roads, and copying records in courthouse basements, and who read copy, made helpful suggestions, and was unfailingly patient, understanding, and enthusiastic.

No book of this kind represents a finished task. Much more remains to be learned. There is always the hope that somewhere, somehow, eighteenth-century records of the old iron furnaces will be found. The final days of Charles Read remain obscure, and the Etna Furnace mystery is not wholly solved. But there comes a time to go to press. As a thirteenth-century Chinese philosopher, Tai T'ung, put it: "Were I to await perfection, my book would never be finished, so I have made shift to collect the fruits of my labors as I find them."

ARTHUR D. PIERCE

Medford Lakes, N. J., 1957

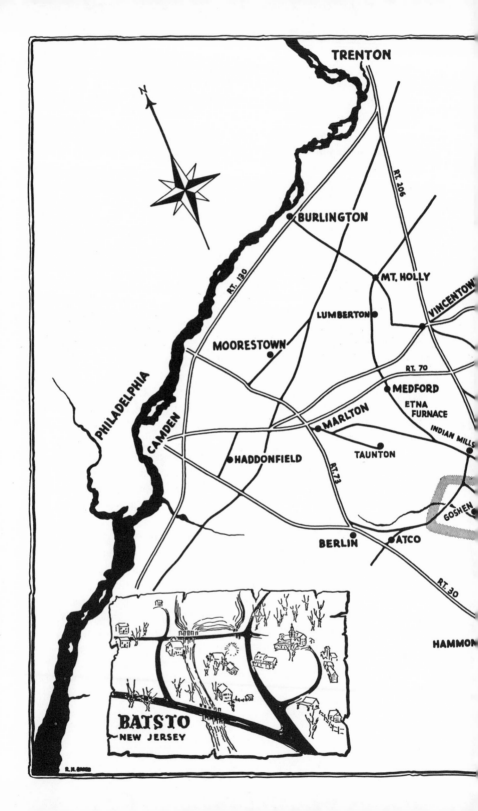

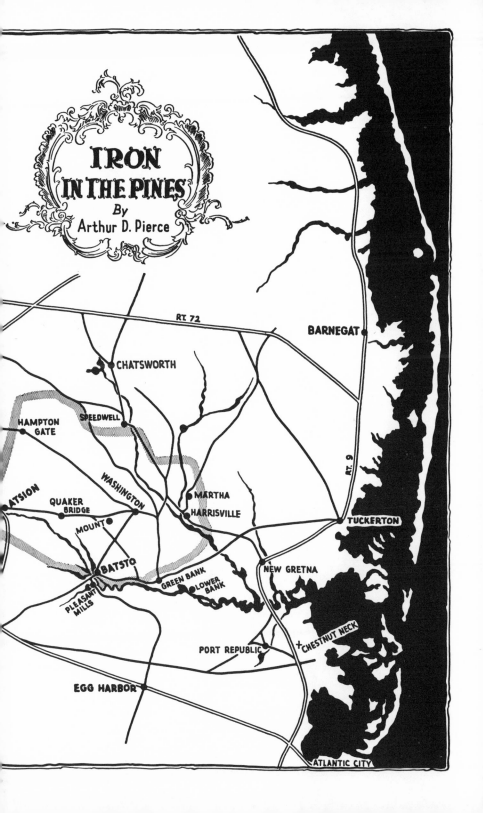

IRON IN THE PINES

1
THE "WHARTON ESTATE"

Few areas in the United States are so heavily travelled, yet so little known, as southern New Jersey. From colonial times men have been crossing the Garden State to bask and bathe on its Atlantic beaches. The count of such journeyers now runs into many millions annually, while a stone's toss from the concrete roads and parkways the wayfarer may find a wooded region primitive in its desolation, haunting in its effect upon the imagination, and historic no matter how neglected the history.

This strange and mighty woodland lies roughly halfway between Philadelphia and Atlantic City, and is commonly known as "The Pine Barrens," although it is not barren, nor are its forests principally of pine trees. The origin of the name is in dispute; but being ancient, colorful, and to some degree appropriate, it has stuck. Now the State of New Jersey has acquired the heart of this forest —a sprawling 100,000-acre tract—for a watershed, game preserve, and state park. For half a century these particular acres have been known as the "Wharton Estate." They extend over three counties: Burlington, Camden, and Atlantic. Beginning about 1873, the late Joseph Wharton, Philadelphia philanthropist, metallurgist, poet, artist, and founder of the Wharton School at the University of Pennsylvania, began acquiring large tracts in the Pine Barrens. He bought crumbling old paper mills, lake sites, and ghost towns. It was Wharton's idea to dam the many Pine Barrens rivers so as to provide a fresh water supply for the city of Philadelphia. Plans were drawn up, and still exist, but the undertaking was blocked when the New Jersey Legislature got wind of it and enacted a measure prohibiting export of water from the state. Thereafter Joseph Wharton concentrated on agriculture, lumbering, and cattle breeding.

The "Wharton Estate" had a remarkable history long before it worked its charm upon Joseph Wharton. Back in the Miocene period, geologists believe, the Pine Barrens were an island in the Atlantic Ocean, at a time when waves were breaking on what now is the Philadelphia waterfront. The pinelands were an island roughly eighty miles long, measured from north to south and thirty miles wide from east to west. Its geologic makeup approximated what is now called the Beacon Hill Formation, and it rose from the sea, according to scientific estimates, more than two million years ago. For many centuries it remained an island; not until much later did the rest of what is now central and southern New Jersey also rise from the sea, to surround it with new land and produce approximately the geography of our own time.[1]

During those many centuries as an island the Pine Barrens developed special characteristics, and they differed sharply as to soil and vegetation from those of the mainland nearby. These differences persist today, and often the contrast is dramatic. On one side of a stream may lie typical Jersey farmland; on the other the soil suddenly ceases to be fertile and, instead, is sand, often as white and friable as if it still lay on the shore of the sea. Open fields give way to forests of pitch pine and oak, tangled with laurel and holly, which are broken by extensive bogs and swamps of white cedar. Even the character of the streams is different. Until they meet the tides, Pine Barrens rivers are of "cedar water," the color of brandy and silky-soft to the skin. And deep in these stream beds may still be found the bog ore upon which an entire iron industry was based in colonial times.

Lush and even exciting in its own way is the vegetation of this extraordinary region. For a century and a half botanists have been coming from as far off as Europe to study the unusual plant life. It was back in 1805—or 1808, the date being in dispute—that the famous botanist Alexander Pursh and three companions discovered the curly fern (*schizaea pusilla*) at Quaker Bridge. Many unusual mosses grow there, and there is an abundance of the famous pixie moss and the sphagnum moss long gathered for use by florists.

Drug plants are numerous and include wild indigo, colic-root, spurge, bearberry, and a wide variety of medicinal herbs.[2] It is commonplace today, as it has been for years past, to find botanists prowling deep in the Barrens, and they now take color photographs as well as gather specimens.

The earliest inhabitants of the "Wharton Estate" were Indians, of a tribe called Lenni Lenape. Not far from Atsion—at Indian Mills, first named Edgepillock, later Brotherton and also Shamong —the first Indian Reservation in the country was established in 1758, and it was there that the Lenni Lenape lived out their last days in New Jersey. White men began arriving in force in the early 1700's. First to come were the woodcutters. Attracted by the water-power which was to lure others in later years, they erected sawmills in large numbers. Soon the devastation of the forests was so great as to cause alarm even in those days of general indifference to problems of conservation. It was Benjamin Franklin who in 1749 urgently advocated conservation and intelligent forestry to combat what was described as reckless and wanton slaughter of the woods.

Next in line came the men of iron, who had discovered bog ore in the stream beds and hastened to exploit it, men like Charles Read, Isaac Potts, Colonel John Cox, Joseph Ball, and William, Jesse, and Samuel Richards. When, in the mid-1800's, the iron industry crumbled to ruin, paper mills sprang up, then glass factories and other enterprises—until the advent of the railroads put an end to it all and turned the clock back in the Pine Barrens, leaving the aged towns to die, the old roads to become overgrown, the forests to thicken, and the streams to flow on in untroubled grandeur.

It is difficult for the casual wayfarer nowadays to believe that the "Wharton Estate" was an industrial region over a century ago; that it was a munitions center for Washington's armies in the Revolutionary War; that the whole area was dotted with sizeable communities built around ironworks, paper mills, sawmills, glass factories, brickmaking establishments, cotton mills, and gristmills; and that ocean-going vessels sailed right into the center of it, anchoring within a mile of the biggest town of all, Batsto. Crossing the area was the Clamtown—or Tuckerton—Road, which not only linked Philadelphia by stagecoach with the seacoast, but served as a main artery for travel between such venerable towns as Atsion,

Quaker Bridge, Washington, McCartyville, Martha Furnace, Batsto, Pleasant Mills, and many more. Some of these communities are today phantoms, but others—Atsion and Batsto particularly—stand much as they did a century ago, and keep alive memories of the old days in the State of New Jersey's 100,000-acre wonderland.

It was after finis had been written to the industrial collapse in the Pine Barrens, that Joseph Wharton moved in. Having purchased the big Batsto plantation in 1876, at a Masters Sale, following foreclosure proceedings, Wharton engaged General Elias Wright as his manager. It was Wright who helped him gather in tract after tract, until the total ran from 100,000 to 125,000 acres. Wharton was fascinated by this region, as so many had been before him. He spent substantial sums upon improvements and experiments. It is estimated that he laid out more than $40,000 on the Batsto mansion alone, adding, in particular, the 86-foot tower and the water tank which is still inside it. Wharton spent some time at Batsto and took a personal interest in his various enterprises, which included, at one time, a venture in raising sugar beets. Cranberry culture was expanded and thrives there today, the huge cotton mill at Atsion having been transformed into a modern, efficient cranberry-processing plant.

Joseph Wharton died in 1909, at 83. Shortly afterward, his heirs offered the entire property to the State of New Jersey, which then as now was much interested in it as a watershed. Called the "last great untapped source of good water within the State of New Jersey," the "Wharton Estate" watershed, according to engineers' estimates, held a potential supply of at least 300 million gallons of water a day. One estimate put the figure as high as a billion gallons a day. The price for the tract in 1912 was $1,000,000, and state officials moved to acquire it. Efforts to conclude a purchase, however, met with opposition, culminating in litigation. Finally it was decided that a state-wide referendum would be necessary to make the transaction lawful under the statutes in effect at that time.

The question of purchasing the "Wharton Estate" as a watershed was put on the ballot in the general election of November 2, 1915. Surprisingly, the proposition was defeated, by a vote of 123,-995 opposed to 103,456 in favor. Even more surprising was the fact that such North Jersey counties as Essex, Passaic, and Hudson favored the purchase, while most South Jersey counties, which presumably had more to gain from the proposal, voted strongly in opposition.[3]

Much of South Jersey's opposition, especially in Burlington County, probably stemmed from the prospective loss of tax ratables. Nine townships and three counties—Burlington, Atlantic, and Camden—were directly affected. Forty per cent of the revenues of one township had come from taxes paid by the Wharton heirs. Other townships faced similar financial difficulties.

Nearly forty years passed. Shortly after World War II the "Wharton Estate" was considered as a site for a United States Air Force Academy. Later there were proposals to have the Air Force purchase it for an overseas freight terminal. The State of New Jersey, however, had not lost interest. To the contrary, as state water problems became steadily more acute, the Wharton tract with its 300-million-gallons-a-day potential seemed ever more desirable and necessary. In 1954, during the administration of Governor Alfred E. Driscoll, the sum of $2,000,000 was appropriated toward purchase of the "Wharton Estate" as a watershed. Meanwhile an appraisal had been undertaken for the state by Kenneth Fisk, Walter D. Lamon, and the late Samuel Walker. A valuation of "just over $3,000,000" was set, and negotiations proceeded.

Agreement was reached on the following terms: $2,000,000 would be paid for about 56,000 acres constituting the eastern portion of the tract, including Batsto, Harrisville, and Martha; also, the state would be given an option to buy the balance of the property—the western 40,000 or more acres, including Atsion—for an additional $1,000,000. The deed for the eastern sector was delivered to the State of New Jersey, and title passed, on December 30, 1954. Transfer of the remaining portion took place the following year, on September 29, under the administration of Governor Robert B. Meyner.

At the time of full acquisition by the state, the Wharton Tract comprised about 2.5 per cent of the total land area of New Jersey. It included the two towns of Batsto and Atsion, their two mansions and 55 other dwellings; the romantic ruins at Harrisville; the lake and slag pile at Martha, and ghost towns—still to be found on maps—such as Quaker Bridge, Washington, and Calico. Included, also, were numerous summer cottages beside Atsion Lake and along the Mullica River. Thrown in for good measure was the old post road—the Clamtown, or Tuckerton, trail—with its memories of fabulous taverns, and its sand-track which still winds gracefully through the forests and which, with care to avoid wrong turns and boggy stretches, may be followed over most of the stage route of yore.

New Jersey acquired some challenging title problems along with the "Wharton Estate." From early times some of the land titles in this area had been none too clear, and early Batsto deeds hint at more than a little overlapping of properties, due to carelessness, or worse, in certain surveys. Because Joseph Wharton assembled his "estate" over many years, such problems tended to multiply. Kenneth Fisk, one of the State appraisers [4] tells how

> one parcel of 1,000 acres was purchased by Mr. Wharton under a deed which contained 21 exceptions. These exceptions range in size from one acre to over 100 acres; they total about 550 acres. The difficulties with this tract start with the fact that the 1000-acre piece is given no metes and bounds but is simply identified as being in the N.W. sector of another and larger section. The parcels which were sold off, or given away, were not identified except as "one acre, part of a tract of—" and so on. In time it may be established that 350 acres were bought by Wharton but he probably occupied the entire 1000 acres as the owners had doubtless forgotten they existed.

Fisk tells an amusing tale of an effort to obtain information from the assessor in a town with a population of about sixty people. He writes:

> The assessor was picking blueberries in the field in back of his house. Asked about Wharton, the old fellow almost dropped his basket: "Those people," he roared, "they claim they own my farm that me and my pappy and grandpappy before him have been paying taxes on for near 100 years."

(It developed that the Whartons probably had clear title but had never taken possession.)

Over most of the "Wharton Estate" there is today a vast and enveloping calm, a sense of peace, relaxation, and contentment. One may be but a mile from the crowded modern highways linking the seashore with the cities, yet in other respects in a world apart, a world without electric power, without telephone or other modern communication. Unbroken wilderness stretches for distances almost unbelievable in a locality less than an hour's drive from Philadelphia and but ninety minutes from Manhattan. It is possible to wander over the side trails for miles without encountering another human being, or to travel by canoe along the earliest and noblest of all the Pine Barrens highways, the rivers, through scenes of rare natural beauty. It is indeed another world, a world which, restored and guarded by the state, will offer pleasure and retreat to many generations to come.

2
THE FURNACES
IN THE FOREST

History in the Pine Barrens really begins with iron—a special kind of iron, bog iron, found in the many streams which flow through the woods and swamps of the "Wharton Estate." Bog iron was abundant there in colonial times; it is equally plentiful today, though no one wants it now. The Indians had made use of bog ore without knowing what it was. In the creek beds they found reddish deposits, and those which were friable they mixed with bear's grease to make paint for their faces. White men soon were to gather those reddish deposits to make fortunes.

Bog ore was first found in New Jersey near Shrewsbury, in Monmouth County, and the Tinton Falls Iron Works was established there about 1675. This was the state's first known venture in ironmaking. Records indicate that it was built and operated with the help of experienced ironworkers who had migrated from Saugus, Massachusetts, after they found themselves jobless when the famous Hammersmith works there closed down.[1] The Tinton Falls enterprise, however, had its own troubles, and many years were to pass before the South Jersey iron industry received a double impetus in the demand for iron created by the French and Indian War and the Revolution, and the discovery of extensive bog-ore deposits in the basins of the Mullica and Wading rivers.

Bog ore is formed by a complex process carried out by nature on a leisurely schedule—in simplified terms, the chemical action of decayed vegetable matter in the streams upon iron salts in the stream beds. Pine Barrens soil is often underlain by strata of marls or greensands which contain a soluble form of iron. The waters of the Pine Barrens rivers and bogs accumulate a high vegetable content. In the many sluggish areas these waters percolate through the marl beds. This produces a chemical reaction which picks up the iron—in solution—and carries it to the surface of the water. There it oxydizes. The resulting oxide is deposited along the banks of the streams, in the beds of swamps and pools, and in the many coves. As the deposits pile up, usually mixed with mud, they harden into thick and rocky ore beds. These are the beds which are "mined" and which yield the substance from which the old furnaces obtained their iron.

Chemically, bog ore is a variety of limonite. According to Charles Boyer, "It is estimated that an exhausted ore bed will renew itself in about twenty years, providing the soil has not been drained and there is sufficient vegetation to act on the iron salts." Apparently these latter conditions did not exist in such a degree as to keep the pineland ore beds from falling behind the production pace of the furnaces, for it is known that Batsto Furnace, and probably others, imported ores from outside the state in efforts to keep going in their later years.[2]

The Pine Barrens possessed an abundance of two other principal requirements for iron manufacture: power, from the streams, and fuel, from the forests. That is why so many furnaces were located in such remote places. The rivers which furnished the ore could, by damming, be made to supply the power required by the smelting machinery. Equally close at hand, the great forests offered a source of charcoal, which was the fuel used to produce iron until well into the nineteenth century. Even a source of the flux, or reducing agent, was close at hand in the oyster and clam shells of the South Jersey seacoast.

The heart of a pineland ironworks was the furnace. A furnace extracted the iron from the bog ore, and the product was often in a rough and porous state. Usually it was given the form of long bars, called "pigs." The story goes that an early ironmaster once compared his furnace hearth and its products to a sow and her pigs. Apt or not, the name has survived. The process of making pig iron was rugged. A furnace operated twenty-four hours a day, seven days a week, all the time it was "in blast." Usually it was "in blast" from seven to nine months of the year. Many furnaces never shut down— barring accidents—until winter froze the streams, halted the water wheels, and stalled the machinery.

Not one of South Jersey's old bog-iron furnaces remains standing today. It is known, however, that generally their outward appearance resembled a pyramid with the top cut off. The hearth was enclosed by a large shed called a "molding house." Because this was a frame structure, it frequently burned down, especially in dry months, when sparks from the stack fell upon the roof. The outer walls of a Jersey furnace were built of native stone, and almost any kind would do. Insulating them from the inner fiery chamber were layers of mortar, brick, and sand. The fiery chamber itself was lined with firebrick, slate, or some other refractory material. Weymouth Furnace, for example, used slate. Remnants of firebrick have been found at Batsto. In shape, a furnace interior was like a bottle open at both ends and flaring out in the middle to from six to nine feet. This widest section was called the "bosh." It was constructed so as to support the burning mixture of ore, charcoal, and flux while the furnace was in operation. Below the bosh was the tuyère, a nozzle through which the air blast was forced. Beneath that lay the crucible, a circular pit at the bottom of the furnace. There the molten metal collected, ready to be drawn off.

The air blast for older furnaces was provided by bellows, great leather affairs with much the shape of those small bellows found by modern firesides. Operating in pairs, these were forced open and shut by means of cams, which were on a shaft propelled by the water wheel. Later on, however, big cylindrical tubs were used at most South Jersey furnaces, as they were elsewhere. There were two pairs of these tubs, each pair operating like a cylinder and piston. One tub fitted closely inside the other, and the inner tub was provided with a leather washer to make a "seal" as it moved up and down inside the larger tub. The ascent of that inner tub sucked in air, and its descent forced that air out through long metal pipes to the tuyère of the furnace. The two sets of these tubs operated alternately, so as to provide a steadier blast. In some furnaces more than one tuyère was used to whip up the terrific heat desired for the smelting process, but available data indicate that most South Jersey furnaces got along with one.

To summarize, this is what went on at a furnace: First, batches of ore, flux, and charcoal were weighed out in given proportions. In many works it required two and a half tons of ore and 180 bushels of charcoal to produce one ton of iron. The ore, flux, and charcoal were taken in wheelbarrows over a trestle bridge to a platform around the top of the furnace stack itself. There they were dumped into the top of the stack, in layers. One filling of the furnace was known as the "charge." Since a furnace usually was in constant operation, this weighing and dumping—by "fillermen"—went on day and night, unceasingly.

Below, in the furnace chamber, the fire was fanned to terrific temperatures by the air blast. This soon reduced the "charge" to a molten mass. Due to chemical action and the effect of the flux, the iron itself was separated from the mass and collected at the bottom in the crucible. The remaining mixture—the slag—was lighter, floated on top, and was drawn off through an opening. When the molten iron was released from the crucible through another opening, it poured out over the hearth in an almost blinding, luminous cascade. There it was guided into long channels in the sand to form pigs, or was ladled into molds. The fluid metal glowed, cooled, hardened, and became solid iron. Even under primitive conditions, this elemental process must have imbued the native vulcans with a sense of awe. The forces employed, and released, were terrifying in their intensity.

Such was the basic furnace. The forge was its more refined cousin. Frequently an ironworks possessed both. Frequently, too, they were close together, although at Batsto they were half a mile apart, and the distance was sometimes even greater. At the forge the often brittle, impure, somewhat porous pig iron was refined and resmelted into malleable or wrought iron, generally called "bar iron." Where pig iron was limited in practical use to such things as stoves, hollow ware, kettles, sash weights, and firebacks (iron plates used for the backs of fireplaces), wrought iron could be used for tools, horseshoes, wagon tires and many other products requiring metal of great strength.

A forge—the operation of which was not unlike that of a blacksmith's forge—changed pig iron to wrought iron by reheating the metal in a blast fire. The molten, pasty mass which was thus created was pounded by heavy hammers, operated by the nearby waterpower. This process removed the excess carbon and other impurities, after which the molten mass was worked into bars or "anconies." The forge hammers, usually weighing more than five hundred pounds apiece, made a great pounding day and night, sometimes jarring the earth itself. While old-timers somehow got used to this monstrous metronome, visitors were apt to find it fearsome and ominous after the normal hushed peace of the forest.

Certain furnaces were so constructed as to combine the functions of furnace and forge. Originally such furnaces were called "bloomaries," but that word came to be so widely misused over the years that its real meaning was destroyed.[3] It was often applied to furnaces and forges indiscriminately. Some iron plantations also had "rolling mills," to make sheet iron, and "slitting mills," which were equipped with mechanical shears to fabricate iron rods, nails, tires for wagon wheels, and similar strip-metal products. Theophile Cazenove, a Frenchman who toured New Jersey in 1794 and visited the Boonton Iron Works, described the latter process as follows:

> The forge for making iron bars is double; a fire and 2 hammers. Bellows [are] of new construction, kinds of iron boilers whose lids are pushed by pistons up to the further end, and from there the air passes through tin pipes into an iron pipe which conducts the air into the fire. In another workshop the bars are made red hot and pass through a roller that flattens them and from there they pass through another roller where the plates are cut into rods suitable to make nails.

He added that the rods sold for 42 pounds, or $105, for a barrel containing 2,000 pounds. Yet another part of an ironworks was the "stamping mill." Equipped with enormous hammers, again operated by waterpower, the "stamping mill" crushed the bog ore before it was mixed into the furnace charge. Also it was used to crush the slag, which frequently contained enough recoverable iron ore to make it worth mixing with fresh ore for another run through the furnace. That so much ore was left in the slag testifies to the low efficiency of those early furnaces. Mention, finally, should be made of the cupola furnace, from which excellent castings could be produced in a single smelting process. In the heyday of their enterprises the Richards family gained a high reputation for castings they made in cupola furnaces by mixing imported Scotch pig iron with the native bog ore. Jesse Richards erected cupola furnaces at both Batsto and the Washington Iron Works.

The comparatively low efficiency of the bog-iron industry spelled its doom. From the start many of the furnaces ran into constant financial difficulties. Some works, such as Batsto, had long periods of prosperity; others passed through a bleak succession of forced sales and sheriff's auctions. Enormous amounts of capital must have been poured into the construction of South Jersey's furnaces and forges, but statistics on the cost of early furnace-building are hard to come by. We do know that as early as 1772 the big Ringwood enterprise in North Jersey cost 54,000 pounds.[4] In 1785 Taunton Furnace, a small one, brought 2,000 pounds at what was more or less a distress sale.[5] In 1830 it cost $230,000 to rebuild the Boonton Iron Works.[6] Gloucester Furnace (near Egg Harbor) sold for $50,000 in 1825, but of that the land value accounted for $35,000. We also know that Colonel John Cox paid 2,350 pounds for Batsto in 1770 and that Joseph Ball gave 55,000 pounds for it nine years later,[7] but the latter sum may well reflect wartime inflation. Clearly it is hard to draw any general conclusions from these figures. Save in the case of Gloucester Furnace, there is no way of telling what proportion of the sums involved represented the extensive land holdings of each furnace and how much the ironworks themselves. That it was a chancy business is attested by the disastrous bankruptcy of Charles Read, the financial collapse of Jacob Downing at Atsion, and the fact that many ironworks were hardly completed before their owners were advertising them for sale, as at Etna, Martha, Speedwell, Batsto, Atsion, and many more.

Resting thus upon a thin financial base, even the best of the South Jersey ironworks were unable to cope with the sharp competition of cheaper—and newer—ways of making iron. Anthracite by the 1840's had proved immensely superior to charcoal as a smelting agent, Pennsylvania's magnetic ores were of purer grade than South Jersey's, and, what was highly important, they were located close to the anthracite fields. The Camden Mail and General Advertiser, of July 1, 1840, noted editorially: "It is suggested that the recent application of anthracite fuel to the smelting of iron ore will be very injurious, if not fatal, to the iron works of New Jersey." That was a prophetic comment. As railroads developed, the bog-iron furnaces often were left isolated and lost any remaining hope of challenging the more strategically located anthracite furnaces. Technological changes were coming, too. Hot blast furnaces were more efficient than cold blast. The little thirty-foot stone stacks gave way to great smelters, soon to tower a hundred feet or more and before very long to be provided with electric power. This in turn was to bring a procession of automatic devices which today are so numerous and so complex as to beggar description. Anyone who has seen the operations of a modern steel plant soon becomes aware that South Jersey's old bog-iron furnaces and forges are far older technologically than they are chronologically. Much more than a single century seems to separate their day from our own.

Below will be found a list of the old furnaces of southern New Jersey. Those outside the "Wharton Estate" have been included because they form part of the broad picture of the times, and because their names may prove useful for reference.

IRON WORKS	BUILDER	OPENED	CLOSED
Atsion	Charles Read	1765	1846 *
Bamber (Ferrago)	John Lacey	1810 *	1865 *
Batsto	Charles Read	1766	1858
Bergen (ex-Washington)	Joseph W. Brick	1832	1854
Birmingham (Retreat)	Bolton and Jones	1800 *	1832 *
Butchers	John Lippincott	1808 *	1840's
Budd's Iron Works	Eli Budd	1785 *	1840 *
Bordentown	Thomas Potts	1725	1748–50 *
	Daniel Coxe		
	John Allen		
Cohansie	uncertain	1773 *	pre-1789
Dover	Wm. L. Smith	1809	1868 *
Etna (near Tuckahoe)	Coates and Howell	1816 *	1832
Etna (Medford Lakes)	Charles Read	1766–67	1773
Federal Forge	David Wright	1789	——
Federal Furnace	Caleb Ivins	1795	pre-1855
	John Godfrey		
Gloucester	John Richards	1813	1848
Hampton	Clayton Earl	pre-1795	1850 *
Hanover	Ridgway, Howell, Lacey and Earl	1791–92	1863–64
Lisbon	John Earl	1800 *	pre-1831
Martha	Isaac Potts	1793	1840's
Mary Ann	Benjamin Jones	1827 *	1860's

Mount Holly	Pearson, Stacy, and Burr	1730	1778 (destroyed by British)
New Mills (Pemberton)	John Lacey	1781–87	pre-1811
Phoenix	uncertain	1816–17	in ruins 1855
Speedwell	Benjamin Randolph	1785 *	1839 *
Taunton	Charles Read	1766–67	after 1830
Union	William Cook, Sr.	1800 *	uncertain
Wading River	Isaac Potts	1795	1815
Washington (later Bergen)	Jesse Richards	1814	1817; later rebuilt as Bergen
West Creek (Stafford)	John Lippincott	1797	1838–39
Weymouth	Shoemaker, Robeson, Ashbridge and Paul	1801–02	1862, forge; 1865, furnace

* Indicates only an approximate date

Some of the men who built and operated the furnaces and forges in these New Jersey pinelands appear in the ensuing chapters. Most of them were men of substance. Nevertheless, when we bear in mind the limits to what they themselves regarded as "luxury," when we consider the rigors then associated with such fundamentals as transportation, sanitation, water supply, medical care, and even cooking, when we reflect that the lot of an average citizen was at least ten times as hard as it is now, then we achieve a better understanding of the role that era played in the civic and economic development of our country. We also gain a better appreciation of the blessings of our own day.

New Jersey's iron plantations, resembling in particular those of Pennsylvania, were feudal establishments and often self-sufficient, or nearly so. Workers rarely lived more than a stone's throw from their jobs, and they labored almost without letup, save on a few special days when they went hunting or fishing. National holidays often passed unnoticed in the early 1800's, even Christmas and Independence Day. There were few amusements save imbibing at the nearest tavern. Wives of workers bought everything at the company store, to which they were more than likely to be in debt long before pay day. The workmen's homes were rude structures. Those restored at Hopewell Furnace in Pennsylvania are of native stone, but those at early South Jersey furnaces were frame cabins, and without doubt the cottages surviving at Batsto rank among the best. Cooking was usually done at the kitchen fireplace, which was the principal source of heat in winter. There was little furniture, and what there was had been roughly built from scrap lumber. Bedrooms were apt to be bare, seldom containing mirrors, tables, wardrobes, or even chairs.[8]

By contrast, fantastic tales have been told of some of the old ironmasters, tales of their luxuries, their excesses, their social life which was as far removed from the workers' cabins a few hundred feet away as the pleasures of Paris were from the plantations themselves. Many of those tales have been exaggerated, and most New Jersey ironmasters appear to have been hard workers and keen businessmen even when they did live on a feudal scale.

During the eighteenth century the benefits of education were not widely available upon iron plantations anywhere. The master's children usually were sent outside to academies, and instruction for the offspring of the few key employees was frequently given by the company clerk. By the mid-1800's, however, quite a few of New Jersey's iron towns had schools. Religious instruction was even scarcer in the earlier days, at least outside the home. The little available was provided by itinerant preachers, whose horseback journeys through the Pine Barrens seem almost legendary now. None the less, these men—men such as John Brainerd and Philip Fithian—laid the foundations, and their pioneer efforts are commemorated by such little frame churches as those which survive in Atsion and Pleasant Mills.

With the collapse of the plantation system, after the furnaces and forges closed down and many of the collateral industries became bankrupt, economic conditions grew even more difficult for the families who had long looked to the ironmasters for security and to the "Big House" and company store for their every need. While paper mills, cotton mills, and some of the sawmills survived, and several glass factories had been established, the old, assured continuity of employment was gone. Often there were severe and unexpected periods of unemployment and poverty. Some of the newer factories in the forest failed. Others burned down and were not rebuilt. Thus, as years passed, more and more families moved away when their breadwinners found jobs in other localities, and the old iron towns became gray shadows of their former selves.

Ruthless as this period seems in retrospect, it was, for some, one of economic opportunity. There are "success stories" of boys born at the iron plantations. Joseph Fralinger, who blew glass as a child at Batsto, was the famous originator of Atlantic City's "salt water taffy." Alfred Adams, born at Martha Furnace, became a wealthy builder and one of the foremost citizens of that same "Playground of the World." [9] For others, all the same, there was grinding poverty, and of the South Jersey iron industry itself nothing was left but memory. Rarely had an industry been so patiently built, and rarely had one been obliterated so swiftly. It was, of course, inevitable. But that substitution of engulfing wilderness for thriving industrial communities always exerts the greatest impact upon those who are making their first acquaintance with what today is New Jersey's "Wharton Estate."

3
EMPIRE IN IRON
The Story of Charles Read

The beginning of major industrial development in New Jersey's Pine Barrens is the story of late autumn in the life of Charles Read of Burlington. A brilliant, often erratic, enormously ambitious, and politically powerful Jerseyman of pre-Revolutionary days, he has been too little known by later generations in the state which he did so much to develop. No portrait of him is known to exist. For nearly a century there was almost no mention of him in written history. His name was later confused with that of his son, who became a traitor in the Revolution; and even the son's shameful acts were long wrongfully ascribed to an outstanding patriot, Adjutant General Joseph Reed, of General Washington's staff.

The Charles Read who is the subject of this chapter was the third in a direct line of five bearing his name. His grandfather, the first Charles Read, was a Quaker, who emigrated from Cornwall at the age of 28, arrived in Burlington about 1669, and later moved to Philadelphia. The second Charles Read became a prominent Philadelphia merchant and shopkeeper, was active in political affairs as alderman, sheriff, excise collector, and judge of admiralty, and in 1726–27 was Mayor of the Quaker City. The third Charles Read was born on February 1, 1715, in a dwelling which later became the famous London Coffee House and which was located at Front and High (now Market) Streets in Philadelphia. His mother —nee Anne Bond— was the second of his father's three wives. His

baptism, when he was twenty days old, took place in historic
Christ Church, where his father was a vestryman.[1]

Young Charles Read had many things in common with his father.
Both father and son were friends of Benjamin Franklin, and customers as well. The elder Read's long career in public office undoubtedly had great influence on the son, as did one of the father's
industrial ventures—a partnership in a company to build an ironworks in Bucks County, Pennsylvania. Both men ended their
careers in financial distress.

Charles spent much of his childhood behind the counters of his
father's busy store. For the sake of the boy's education, his father
sent him to London, where he stayed with influential friends of
the family. After several years he became a midshipman in the
British Navy and was assigned to a man-of-war, the *Penzance*. As
part of the West Indies squadron, the *Penzance* operated out of
the British naval base at the island of Antigua. There young
Charles's career took an important turn. In St. John, capital of
Antigua, he was frequently entertained by the family of Jacob
Thibou, whose father had spent several years in Philadelphia and
there had become acquainted with the elder Read. Soon the old
ties between the two families were supplemented by new ones, for
here, on this romantic West Indian island, Charles Read wooed
and won Alice Thibou. Since he never had liked the sea, he then
gave up his Navy commission without hesitation or regret.

Jacob Thibou was an important man in Antigua. A planter possessed of wealth and power, he was also Chief Justice of the An-
tiguan courts. His family included two sons and ten daughters.
Alice, the sixth child, was then 18. Aaron Leaming, Charles Read's
sole contemporary biographer, and scarcely a friendly one, describes Alice thus: "not hansom, nor gentele, but talked after the
Creole accent." It has been suggested that the marriage was one
of convenience, since three months before, Charles's father had
died in Philadelphia, intestate and heavily in debt. Down in Antigua, however, according to Leaming, Charles "passed for a rising
genteel young fellow the son of a rich merchant and eminent
grandee in Philadelphia."

The marriage of Alice Thibou and Charles Read took place on
April 11, 1737, at St. John's. When the young couple set off for
Philadelphia by ship, the father of the bride sent along an alcoholic

dowry; he "ordered the negroes to rool out 37 hogsheads of rum."
Many more, says Leaming, were "consigned to him" in Philadelphia, so that Charles Read "made his appearance . . . in quality
of a rich merchant."

Perhaps Philadelphia society was cool to Read, as a result of his
father's financial failure. It may be that Alice Read found herself
uncomfortable in the confined Quaker circles of the day. In any
case, after two years the Reads moved to Burlington, New Jersey.
There, in 1739, Charles set out to build a career and seek a fortune. His start was modest—a job as Court Clerk of Burlington
County, with a salary of 60 pounds a year. Yet Read became, in
about a decade, the most politically powerful man—some called
him the dictator—in New Jersey. He far surpassed his father in the
matter of holding public office. In his swift ascent of the political
ladder he was at various times—often combining several roles—
Clerk of the Circuits, Surrogate of the Prerogative Court, Indian
Commissioner, Deputy Secretary of the Province, Member of the
Assembly, Member of the Governor's Council, Associate Justice
of the Supreme Court, and, for a short period, Chief Justice of
that tribunal. Read had been admitted to the bar without a formal
legal education, but he soon built up a large practice. He was the
close confidant of three Governors—Morris, Belcher, and, for a
time, William Franklin, the son of Benjamin. For nearly thirty
years Read was the chief dispenser of political patronage in New
Jersey, and he was accused of making the most of that power.
Aaron Leaming said of him: "No man knew so well as he how to
riggle himself into office, nor keep it so long, nor make so much
of it."

How Read found the energy, let alone the time, to assume so
many public obligations is difficult to understand. Certainly he
took his tasks seriously, and he would seem to have deserved a
more sympathetic biographer than Leaming. Read's accomplish-
ments as Indian Commissioner were extensive and important and
included establishment of the first Indian Reservation in the
country at Edgepillock, Burlington County. As a legislator,
Read had a phenomenal record. He led the way in solving the
frequently vexing financial and military problems of the Province;
he promoted road and waterways improvements, timber conservation, and agriculture, while among his many less spectacular

achievements was a statute which provided for the preservation of New Jersey's public records.

Aside from his official duties, Read was busy with enterprises on his own account. He speculated heavily in undeveloped lands, although he does not appear to have profited greatly from such ventures. He was not only a farmer, but a genuine student of agricultural science, whose voluminous notes still command respect and have been happily resurrected in Carl R. Woodward's biography of Read, *Ploughs and Politicks* (published in 1941, but long out of print). Woodward notes that Read's "farm notebook . . . is one of our richest known sources of information about farming in the American Colonies."

At the age of 51, Charles Read set out to make a new dream come true—his dream of becoming the greatest ironmaster in New Jersey. A measure of his enthusiasm for this venture is found in a letter from his cousin, William Logan, of Philadelphia, to his brother, James Read, of Reading, Pennsylvania. Dated February 4, 1766, it says in part: "I spent this evening with thy Bro. Charles who is just come down to Burlington to stay a day or two. . . . He is quite hearty and very full of spirits and his iron works scheme." [2] Read certainly tackled his "iron works scheme" with characteristic vigor and thoroughness. He had the metallic content of bog-iron deposits carefully analyzed. Through a maze of conveyances, agreements, and assignments he acquired the thousands of acres necessary to his projects and purchased the rights to dig iron ore and cut timber in thousands more acres to which he did not take title. In addition, he negotiated the rights to dam streams, obtained special legislation to advance his enterprises, and somehow carried on his various public offices and his law practice at the same time.

Read built Etna, Taunton, Atsion, and Batso in rapid succession, and his whole chain of forges and furnaces was completed and in operation by 1768. To his apparent surprise—Read seems not to have kept careful track of money—his "iron works scheme" was consuming more capital than he had anticipated and more than he was able to provide. Even before the projects were fully launched, he was advertising for prospective investors. Most of those whom he found were already friends and political associates, but even with their help he was hard pressed. Batsto Furnace was

sold almost as soon as it was opened, and Read had three partners
there from the beginning. A 449/1000 interest in Atsion was sold
in 1768, and in 1770 Read advertised to offer all his interest in
Etna, Taunton, and Atsion. His controlling share in Atsion was
finally sold in 1773, together with the Etna machinery, but for
Taunton there were no takers.

Not only did his ironworks exhaust Read's capital resources and
credit, they taxed his health and his energies far beyond capacity.
It became physically impossible for him to give them the personal
attention they needed and also carry on his still numerous official
duties and his law practice. As he wrote later, "My Constitution
would not enable me to look after such an Extensive business
. . . & most of the people I employed deceived me." Meanwhile
other blows were falling. Read began to suffer from recurring ill-
ness, which frequently kept him from attending sessions of the
Governor's Council. He had borrowed more and more, including
500 pounds from his cousin, William Logan, secured by a mortgage on Etna Furnace, and later another 40 pounds from the same
source "for hammers" at Etna.[3] Poor health also plagued Alice
Read, and it was a major shock to her husband when death finally
came to her at their home in Burlington, on November 13, 1769.
Her burial, in St. Mary's Episcopal Cemetery in Burlington, was
attended by the Chief Justice of New Jersey, by the Attorney
General, "Gentlemen of His Majesty's Council," and a host of other
notables, a tribute which attested her husband's position and
stature. Alice Read was only 51 when she died, and Charles himself was but 55 the following year, when, in offering his three
remaining ironworks for sale, he stated: "The only reason for selling them is that it is necessary to have a person concerned in the
works resident at Philadelphia and a man of activity at the furnace.
The present owner is very infirm and not able to stirr much."

After Alice Read's death, Charles went to live with his son at
Etna Furnace. There misfortune continued to beset him. During
the winter of 1770–71 he was so ill that Charles, Jr., had to carry
on his correspondence and on January 23, 1771, wrote that his
father "has violent inflammation in his left leg where a wound he
had about twelve years ago has broke out." In a letter dated the
following May, Read himself wrote, "My recovery [was] not expected by my friends." Of course, he did recover, but it is doubt-

ful that he was very happy at Etna Furnace, for he later referred
to "the unrelenting spirit of country people" and their "itch for
tattling and reports without Foundation," which gave him "great
uneasiness." Further grief came two winters later with the death
of his "dear little grand daughter . . . of the measells . . . her
brother went through them with great difficulty but is now toler-
ably well." Finally, aside from such troubles, winters in the country
were rigorous in those days and taxed Read's failing strength.

The melancholy affairs of Charles Read reached their grim
climax in the year 1773. While Alice Read had left a considerable
estate, most of it was in Antigua and could not be liquidated in
time to stem the tide of her husband's financial reverses. Pressed
by his creditors "to gett my son bound with me," Read felt that
"it was time to be out of the Way." With great secrecy, he prepared for flight. On April 3, 1773, he deeded Etna Furnace with
about 8,000 acres of land to Charles, Jr.,

> in consideration of a Debt due from him the said Charles Read,
> Esq. . . . for his Wages due to him as Manager at his Iron
> Works amounting the 25th day of May 1768 to the sum of four
> hundred and twenty pounds and for which the said Charles
> Read the Elder executed on that day a Bond for the payment of
> the same with lawful interest.[4]

Part of this transaction provided for repayment to William Logan
of the 540 pounds borrowed from him. Significantly, while the
deed was executed on April 1, 1773, it was not presented for
recording until June 28, 1773—long after Charles Read had left
the Province of New Jersey far behind him.

The first public knowledge of Read's departure came from a
newspaper notice in *The Pennsylvania Gazette* of June 30, 1773:

> WHEREAS CHARLES READ, Esq. for the recovery of his
> health, as well as for securing and recovering some large sums of
> money due to him in the West-Indies, has lately embarked
> thither, and, being desirous of preventing any uneasiness among
> such as he may owe money to, has appointed us, the subscribers,
> trustees to make sale of such parts of his estate, as may be neces-
> sary for the discharge of his debts, which we propose to do as
> soon as possible. We, therefore, desire all persons who have any

demands against him to bring in their accounts, properly proved, that they may be settled; and all who are indebted to him, by mortgage, bond, note of book-debt, are desired immediately to discharge their respective debts to the subscribers, who are authorized to receive the same.

Daniel Ellis, at Burlington
Charles Read, Junior, Aetna Furnace
Thomas Fisher, Philadelphia

Not only did Read leave New Jersey secretly, he never filed any resignation from his various public offices. Quite possibly, he hoped to obtain enough from his wife's estate in Antigua to enable him to return, and in a letter from St. Croix, in the Virgin Islands, to his trustee, Daniel Ellis, he wrote: "I hope no man shall lose by me tho' they may be longer than I choose out of ye money."

Deep mystery has long surrounded this tragic flight of Charles Read. Apparently he went first to Antigua, where his hopes concerning his wife's estate were not realized. Then he went to St. Croix, where he lived for some months. It was the good fortune of the author to come upon a letter of Charles Read to Israel Pemberton, dated December 15, 1773, which had not caught the attention of previous historians and which, after so many years, reveals the emotions and motives impelling a man of such stature and influence to leave home, kin, and friends and put behind him almost everything which had made his life worthwhile. It has been suggested that Read was of "unsettled mind." This warm, human, and moving letter seems to belie any such supposition. The pertinent portions of the letter follow:

St. Croix, December 15, 1773

I wish I could bring myself to a resolution of complying with their Solicittations [Presumably friends and family had urged him to return.] but I so farr know myself that the most terrible consequences would attend it. When I first discovered the bad situation of my Affairs it struck me with Amazement & for a long time I was racked to such a degree that death was more Eligible but that the Dictates of Conscience forbad it should have putt an end to my troubles. I did what appeared to me most agreeable to the rules of Equity & never intended nor do I think I have given any undue preference among my creditors. I expected and hope with good management there will prove enough to satisfy every Just debt. If there is not & it be ever in my power It shall be done. It was with intention to gett rid of intolerable torments of mind that I left my home. I had no Fortune nor inclination to attempt the increasing it by Entering into a life of Hurry and bustle. Quiet & Ease is (what) I could not Enjoy at home . . . and when some of my Creditors last Spring pressed me to gett my son Chas bound with me it was time to be out of the Way. I had seen often the unrelenting spirit of country people & dreaded their . . . importunities Altho I believe I had many good friends in New Jersey the itch of the Country people for tattling and reports without Foundation gave me great uneasiness.

It cannot be said by the most malicious that I wasted my Substance in riot or on Indulgence in any Expensive vice. My Constitution would not enable me to look after such an Extensive business. My son disliked it & most of the people I employed deceived me. I fell a sacrifice to my Ignorance of & Inattention to Accounts. I could not bear the Severity of Winter and my Confinement at that Season rendered me listless thro the Year. This determined me to seek a warmer climate where the Excessive price of the necessaries of life (would not) terrify me. . . . I am determined if possible to putt myself in such a situation as to live without hurry and confusion that I may calmly (review) my past life . . . Spin the short thread remaining in comfort and fitt to leave this world. . . . I seem to have no more (business resources) I served New Jersey with great faithfulness & integrity & till lately my services were by no means properly considered. In the Secretary's office I did never Exact the rigour nor would my conscience permit me to take what they now do. . . . [Words in parentheses are indistinct, or omitted, in the manuscript.]

In this document, the character of Charles Read begins to appear in a far better light than that cast by the hostile Aaron Leaming. Leaming called him "high strung and selfish, not very grateful for good offices, unwilling to forgive an injury, not very faithful to his client's cause . . . timerous almost to cowardice . . . whimsical to the borders of insanity which he inherited maternally . . . and he knew no friend but the man that could serve him." While the historian still does not know too much about Charles Read the man, he does know better than that. When the

Province of New Jersey was short of cash, Read willingly went without his salary to help out. His was a steady, guiding influence in public affairs for many years, and, as he put it, his conscience did not permit him "to take what they now do." In contrast to Leaming's comment that "His intrigues with women . . . employed a large share of his thought," he himself asserted, "It cannot be said by the most malicious that I wasted my Substance in riot or on Indulgence in any Expensive vice." On one point, certainly, Leaming seems to have been right—when he said Read "loved the country better than his family." To Leaming that was a fault. To many it would be a virtue. Read served well and faithfully. His mistakes were those of the visionary and leader; and even though his "iron works scheme" seemed like utter folly to his contemporaries, Charles Read built better than he himself ever knew. While he did not live to see the day, his furnaces and forges —notably those at Atsion and Batsto—served as precious arsenals for the Continental armies in the Revolution, and became the nuclei of the substantial bog-iron industry which later developed in the New Jersey Pine Barrens.

Darkness still obscures the final year of Charles Read's life. It is not known just when he left St. Croix, seeking a situation in which he might "live without hurry and confusion." All that is known is that, reverting to the mercantile instincts of his early youth, he went to Martinburg, North Carolina, on the Tar River, and there opened a store, which Leaming called "a small shop of goods." It must have been a strange and lonesome life, and Read surely made a pathetic appearance in that backwoods hamlet. No letter from him at Martinburg has come to light. Perhaps he wrote none. Perhaps by then his mind had given way. Perhaps he found the peace he sought in this utter obscurity. In any case, all was soon over. On December 27, 1774, Charles Read died, alone. Whatever sort of funeral he had, it was unattended by family or friends. No one even knows where he is buried. The only record of his passing is a note in Leaming's diary, for November 14, 1775: "When I was in Burlington Jacob Read informed me that his father The Honourable Charles Read, Esq., died the 27th of December 1774 at Martinburg, on Tar River 20 miles back of Bath town in North Carolina."

Such was the tragic end of a distinguished man and a commanding career. Read did not even leave a will. That fact was to plague his family, who apparently did not even learn of his death until long after it had occurred. Charles, Jr., seems to have gone to Martinburg in 1775 for his father's effects. Even in North Carolina, however, there was litigation over his assets, for in 1776 authority was granted there to sue "all people Indebted to the estate of Charles Read, Esq., decd."

4
ATSION

Atsion is best approached in the afternoon. The western sun gives a fine glint and shimmer to the large lake and breaks through the trees to lend light and an illusion of warmth to the empty old mansion which dominates the scene. Ancient as it appears, with its long verandas, lofty walls, and staring windows, this "Big House" is by no means as old as Atsion itself. Some have said that the place is haunted, and certainly its suggestion of luxurious living in a bygone day, the strange "belfry" on the aged store nearby, and the mysterious music of the winter winds as they bend the cedars and pines, all combine to excite the imagination. Not even the presence of a concrete highway, U.S. Route 206, cutting through the heart of Atsion, and the monotonous rumble of giant trucks going by on that highway have wholly destroyed the spell handed down from the past.

Unlike most Pine Barrens communities of New Jersey's "iron age," Atsion has enjoyed no less than four revivals in the form of local industrial booms, each of which flourished in its turn and then petered out. Since these left a legacy of structures more durable than the frame shanties of such iron towns as Hampton, Martha, and Speedwell, Atsion today consists of a weather-beaten, but still substantial, remnant of the old town on the eastern side of the concrete highway, while on the western side a modern cluster of cottages nestles comfortably around the lake. In all probability there are more dwellings in Atsion today than at any time in its history, but the economic roots of most of their occupants lie elsewhere now, whereas nearly two centuries ago, Atsion itself was a burgeoning industrial center.

The name Atsion comes from the Indians, who called the strong, cedar-colored stream the Atsayunk, or Atsiunc. Running along the southerly line of Shamong Township, the Atsayunk originates some ten miles to the west, near the hamlet once called Long-a-Coming and since 1872 known by the more commonplace name of Berlin. The Atsayunk for much of its course toward Atsion Lake forms the boundary line between Atlantic and Burlington Counties, as signs on little bridges deep in the pinelands take pains to inform the traveller. It was the potential waterpower from this stream, together with rich beds of iron ore in the lake and its adjoining bogs, which attracted the keen interest of Charles Read.

For some years Read had been buying and selling, often without much apparent profit, tracts of land in this section of the Jersey wilderness, while exploiting certain of those tracts by building or leasing sawmills and cutting timber. On March 31, 1755, Read, together with Thomas Gardiner, made his first purchase of land in the immediate area of Atsion—a total of 1,133 acres.[1] Read bought only five acres outright. Title to the remainder vested in Gardiner, who was one of Read's friends in Burlington and from whom, in 1744, he had bought 1,725 acres elsewhere in Burlington County. (In 1749 Read had written a friend "My Estate lays chiefly in Land.") A curious series of transactions followed. On August 9, 1757, Gardiner sold a half interest in his holding to Daniel Ellis, another of Read's Burlington friends. Then, on September 10 of that year, the two men—Gardiner and Ellis—granted Read a 999-year lease on their tract. Payment was fixed at four pounds, ten shillings, "proclamation money."

On June 29, 1765, John Estell obtained permission from the New Jersey Assembly to build a dam across the Atsayunk, the power to be used for a sawmill. It is possible that he was acting for Read, because, shortly thereafter, Read bought Estell's rights and late in 1765 built Atsion Forge on that waterpower.

Construction of Atsion Forge is dated by an agreement between Charles Read, David Ogden, of Newark, and Lawrence Saltar, of Nottingham, Burlington County. This agreement set up the partnership and company which were to operate the new iron enterprise. It was signed on January 9, 1766, and in it Read assigns to his two partners "the one moiety of a tract of land whereon the said Forge is erected and which was purchased of James Inskeep by deed dated the 19th of July, 1765." Making necessary allowance for weather problems in the locality, Atsion Forge probably was erected soon after the date of the purchase from Inskeep, in late summer and early autumn. In any case, it already had been erected before the agreement was signed on January 9, 1766.

This same historic agreement, only recently located, provides for sharing the "expence arising on the building, finishing and furnishing stock for the new Forge called Atsion." It stipulates that "All servants and apprentices belonging to or working at the Forge shall be for the mutual advantage of each owner in proportion to each equal share," and that the "Manager's wages shall not exceed One Hundred Pounds per annum and the Clerk's Forty pounds per annum." There was a further provision that "All Surveys or Purchases to be made shall be in like manner . . . on any land lying to the Northward of the mouth of Sleepy Creek and Eastward of Mechesetauxing and Southward of the Indian Town." [2]

In its early days Atsion Forge was employed chiefly in converting pig iron, brought from Batsto eight miles away, into bar iron. While Read's analyses of the bog ore in Atsion Lake suggest that he planned a furnace also, that was not to come until Read himself was out of the Atsion picture. As noted in the preceding chapter, he had been unwell; and with his official duties and other extensive interests, it had become physically impossible for him to supervise the chain of ironworks he was establishing. One of Read's partners, Lawrence Saltar (sometimes spelled Salter), son of Justice Richard Saltar, bought 249/1000 of Read's leasehold. The other partner, David Ogden, Jr., a member with Read of the Governor's Council, purchased 250/1000. This left Read with the one-thousandth margin of control. Operation of the Atsion works seems to have been assigned to Saltar, the role of Ogden probably having been that of supplying fresh capital for the venture. By 1770 there had been built at Atsion "four forge fires and two hammers," together with the "necessary buildings." [3] It was a sizeable plant for its day. How many dwellings were constructed is a matter for conjecture, but it is known that Saltar made his home at Atsion, at a location which old maps show to have been not far from the site of the present mansion.

That same year, 1770, Charles Read's financial difficulties and physical ailments led him to advertise in the *Pennsylvania Journal* that his interest in the "Atsion Forge or Bloomary" was for sale. This advertisement stated "There are at the works several servants and negroes who understand different branches of the business." Prospective purchasers were slow to appear. Economic conditions were poor in the wake of the French and Indian wars, and with the Revolution soon to come, political conditions already were in ferment. In addition, there was the constant threat from London of legislation to harass the iron industry in America. For years British ironmasters had tried to put through the House of Commons measures for "protection of His Majesty's forests in America," a high-sounding goal which they proposed to achieve by putting the colonial iron-makers out of business.

Not until March 16, 1773, on the eve of his final and tragic flight from New Jersey, was Read able to sell his interest in Atsion. The purchasers were Henry Drinker and Abel James, of Philadelphia. This sale seems to have been part of a general reorganization, for on April 2 of that year Daniel Ogden sold his quarter interest to Lawrence Saltar. Thus, under the new setup, the ownership was divided as follows: Lawrence Saltar, 449/1000; Drinker and James, 501/1000.

Together these partners formed the Atsion Company, which operated the works with apparent success for nearly thirty years. The town grew apace. The first major expansion of the new ownership was the building of a furnace, with the blast machinery purchased from Etna. It was put into operation by 1774. This permitted the Company to become independent of Batsto Furnace, which also had passed out of Read's hands, and further enabled the partners to exploit the wealth of bog ore right at hand in Atsion Lake. Analysis of the Atsion ore had shown it to be from 45 to 47 per cent metallic iron, with some samples showing as much as 56 per cent, and Boyer writes that "Great masses were taken from the bed of the pond during the winter, when the fires in the furnace and forges were out and the water drawn off." Much of it also "was dug three or four miles above the furnace and floated to it on barges," according to K. Braddock-Rogers, who gives the following interesting description of ore-raising there:

The swamps were and still are very widespread and the numerous shallow coves at their edges were covered with water from one to two feet deep. The ore was dug chiefly in these coves. Excavations eight to ten feet square were made and between each a thin dike or dam was left to prevent the water from flowing in upon the diggers or ore raisers.[4]

As Atsion's prosperity increased, three new sawmills were built, and also a gristmill. With an ample supply of labor from the nearby Indian Reservation of Brotherton (now Indian Mills), which Read had been instrumental in establishing, Atsion was soon enjoying prosperity. During the Revolution, however, the Drinkers, being Quakers, shut down the furnace. Saltar, however, continued to operate the forge, which was under his control, and this seems to have marked the first conflict between the partners. [Henry Drinker was one of the Tories confined in Virginia by the Patriots.] In any case, Atsion forge was busy on war orders. No one knows just how much material for the American Revolution was produced at Atsion, but Boyer mentions the probable supply of 170 camp kettles to the Pennsylvania Committee of Safety, and in 1776 the Atsion Company got an order to furnish the Pennsylvania Navy with iron, although what form that iron took—munitions, ship gear, or both—is not known.

Atsion had long been linked with the Delaware River valley by primitive highways: the Tuckerton Road from Cooper's Ferry (Camden); the Shamong trail from Burlington; and subsidiary sand-tracks, winding through woods and fields from Haddonfield, Moorestown, and Long-a-Coming. These roads served also to connect Atsion with Charles Read's other iron ventures at Taunton and Etna (Medford Lakes). Over them one of the earliest recorded pilgrimages was made through the Pine Barrens to Atsion. It was in March of 1773 that Henry Drinker bought his interest in that ironworks, and the following month Elizabeth Drinker, Henry's fashionable though Quaker-born wife, set out with him to see the property for herself. She tells of it in these notes from her Journal:

1773—April 11. First Day. After dinner, H.D., Josey James and myself crossed the river Delaware. I rode on ye old mare as far as Moores-town; I have not been on horseback for 15 years past. Lodged at Jos. Smiths. Next morning we set off in borrowed

waggon with our 2 mares for Atsiunc at the Ironworks, Polly
Smith with us. We stopped at Charles Read's Ironworks 10 miles
from Moorestown, [Etna], then went 10 miles farther to Lawrence
Saltar's where we dined late. Went in ye afternoon to ye forge—
saw them make Bar-iron.

April 16—Left Atsiunc. Lawrence S., his wife and sister with
us; made a short stop at C. Read's [Etna].

By 1804, apparently, the partnership of Drinker and the Saltar
heirs (Lawrence had died in 1783) was in rough water. Again our
informant is Elizabeth Drinker. In her Journal for March 21, 1804,
she relates that Henry Drinker had appointed Reynold Keen, Jr.,
for some years a clerk at the Atsion works, to be "overseer or
manager of ye works. Wm. Saltar took the keys of ye stores from
him by force, and would not let him have a horse on which to
come to town. Very ill conduct, and very ungrateful, all things
considered."

No records are available to show what transpired further, but
the rift either widened, with William Saltar trying to dominate,
or else the works were having money troubles, since the very next
year, 1805, the Atsion property was sold at public auction in the
Merchants Coffee House in Philadelphia. At that time the estate
included a blast furnace, an air furnace "in good repair," a forge
with four fires and two hammers, a sawmill, and two gristmills,
together with the ore beds and approximately 20,000 acres of land.

The buyer of Atsion at the Coffee House auction was Jacob
Downing, son-in-law of Henry Drinker. Actually, Downing owned
but a half interest.[5] A significant side light on the whole transaction is found in an agreement dated December 30, 1808, four
years later, in which Downing declares:

> notwithstanding the whole of the "Atsion Estate" was conveyed
> to me. . . . I claim one-half of the said estate only, the other
> half remaining to be the property of Henry Drinker, the elder,
> and I hereby bind myself, my heirs, executors and administrators
> . . . that whenever I shall be discharged from all responsibility
> respecting my endorsements . . . of certain notes negotiated at
> the Banks in this City [Philadelphia] and that I will execute a
> good and sufficient deed, vesting in the aforesaid Henry Drinker
> . . . the aforesaid one-half of the premises above described.

Under the Downing regime this ironworks in the pines prospered for a number of years. The community had grown substantially in its first four decades, and back in January of 1798 the federal government had considered it of sufficient importance to open a post office there. Surviving evidence of Jacob Downing's tenure is found in several stoves. Two of these are in the First Presbyterian Church of Bridgeton and bear the legend: "Jacob Downing—Atsion Furnace." Additional products of Atsion's early days included pots, pans, kettles of varying sizes, and firebacks molded with pleasing designs; one of the latter now reposes in the Historical Society Museum of Bucks County, Pennsylvania, and another is in the Gilder House in Bordentown, New Jersey. Several other Atsion stoves are known, of varying vintage. One of these still keeps the Quakers warm in their meetinghouse at Crosswicks, and another —a beautiful Franklin Stove—is in the Bucks County Museum.

Deduction suggests the year 1815 as marking the probable beginning of Jacob Downing's financial difficulties, which brought this first era of Atsion to a melancholy close. It was in that year that the federal government moved the post office from Atsion to "nearby at Sooy's Inn," which was actually almost eight miles distant over the Tuckerton stage road, at the place long called Washington.[6] Whether this moving of the post office means that the ironworks were closed down at that time is a matter of conjecture, but it certainly indicates a sharp drop in postal patronage at Atsion.

Two years later, in 1817, Downing mortgaged his interest in the "West Mills Tract" (southeast of Atsion proper) for $12,556, apparently to satisfy four overdue promissory notes. Matters soon grew worse. Downing defaulted on both principal and interest on this mortgage, and under foreclosure proceedings the tract passed to the Bank of North America (which sold it to Samuel Richards on July 10, 1822).[7] Then came further borrowings, and new difficulties. The property was neglected, the ironworks came to be deserted, and by 1823 there was little but desolation in Atsion. That hardy traveller, J.F. Watson, passed through the town in 1823 on his trip to "Longbeach-Seashore," and wrote as follows:

Was much interested to see the formidable ruins of Atsion iron works (27½ miles). They looked as picturesque as the ruins of

abbeys, etc., in pictures. There were dams, forges, furnaces, store-
houses, a dozen houses and lots for the workmen, and the whole
comprising a town; a place once overwhelming the ear with the
din of unceasing, ponderous hammers, or alarming the sight with
fire and smoke, and smutty and sweating Vulcans. Now all is
hushed, no wheels turn, no fires blaze, the houses are unroofed,
and the frames, etc., have fallen down and not a foot of the busy
workmen is seen.[8]

Of the Atsion which existed then, there is nothing left today,
save the mossy and moldy fragments of the foundation of a gristmill, and some slag, buried in the weeds several hundred yards
from the present dam site. There is fascinating hearsay to the effect
that a portion of the base of the old forge is under the present
dam, and that stone from the original buildings was used in part
for its construction. . . . Aside from such doubtful relics, the
Atsion of Charles Read, Henry Drinker, and Jacob Downing is a
misty memory, and as in ancient times an entirely new town was
built upon the rubble and ruin of the old one.

In the year of Watson's journey, 1823, Jacob Downing died. Less
than twelve months later, all of Atsion had passed into the strong
hands of Samuel Richards, whose father, William, founder at
Batsto of the Richards "iron dynasty," often had been entangled
with the Atsion Company in litigation over water and ore rights.
(The deeds to the Atsion property, incidentally, are fascinating
documents, which leave the reader amazed at the way surveyors
of that period would describe the boundaries of a tract in terms of
trees, stumps, creek-bends, and other highly perishable features of
the landscape.) In addition to the original property of Charles
Read, the Samuel Richards estate included the Hampton Furnace
tract to the north—later a flourishing cranberry bog—and the West
Mills tract already mentioned. The total estate amounted to more
than 60,000 acres.

When Samuel Richards purchased Atsion he was 54 years old
and still enjoying that vigorous health characteristic of so many
members of his dynamic family. Six feet tall, of powerful physique,
he had inherited all the forcefulness of his father, William, from
whom he had learned the iron business at Batsto. In 1808 Samuel
had acquired the Weymouth Furnace, and that same year he had
also purchased Martha Furnace in partnership with Joseph Ball.

Thus Atsion's new owner was a skilled and practical ironmaster of long experience, and under his administration the community reached its peak of prosperity. Samuel Richards was born to wealth and twice married wealth. His first wife, Mary M. Smith, was the daughter of a rich Philadelphia shipping magnate long known as "Silver Heels"; he is said to have had bullion hauled to his home by the wagonload. Samuel's second wife, Anna Martin Witherspoon, was the widow of a wealthy New Yorker. The master of Atsion himself said that good luck had much to do with his prosperity, and cited the case of a customer who was unable to pay for his iron and offered some lots in Brooklyn to cover his $9,000 indebtedness. Richards had held them only a few years before they brought him $60,000 under the hammer.[9] Eleven children were born to Samuel Richards, eight by his first wife, three by his second. Five died in childhood. Only one son, Thomas S. Richards, followed his father in the "iron tradition" of the family, although a daughter, Sarah, married Stephen Colwell, long manager at Weymouth Furnace.

Atsion caught Samuel Richards' fancy, and he proceeded to exercise the prerogatives of possession with characteristic vigor and thoroughness. Moving swiftly to reopen the ironworks and rehabilitate the town, he also sought new outlets for Atsion's products. For Philadelphia, he turned out fittings and equipment for its water system, particularly the Fairmount plant. For Trenton, he made castings used in the bridge across the Delaware. Other products, including a wide variety of stoves, were hauled to the "Atsion Wharf" at Lumberton and then shipped by water. The Atsion account books show that by 1826 Richards was engaged in a "large trade with merchants along the Hudson River . . . and a schooner named 'Atsion' plied regularly between the Mullica River and Albany, N.Y. carrying the finished products to New York City, Poughkeepsie, Albany and Troy." [10]

It was in 1826 that Samuel Richards built in Atsion the square mansion which still stands triumphant over time and weather. The old Saltar mansion had fallen into decay, but this new "ironmaster's house" was built to last. Samuel had made his home on Arch Street, above Ninth, in Philadelphia, but after the Atsion house was ready he regularly spent spring and summer there. Con-

temporaries have told of the great housewarming—how Samuel
Richards rode from Philadelphia to Atsion for the event in a grand
coach-and-four, and how this first big party at the house was
"attended by some of the best known men of the day." It was, of
course, only the first of many such social occasions, for which the
mansion was graciously adapted; press accounts tell us, "Richards
left behind him the reputation of being a good-liver, with all that
term implies."

The Atsion mansion was built across from the lake and facing
the old stagecoach and post road, which winds eastward toward
Quaker Bridge and Tuckerton. The visitor passes a still imposing
hedge of English boxwood, then follows a gravelled walk leading
to the veranda and main doorway. Prior to restoration, the walls
of the mansion had been stained yellow for the most part and
overstained red on the first floor. They are of Jersey stone, smooth
cast. The veranda runs the length of the south side of the house,
is supported by square pillars of the same stone, while upon these
pillars, holding up the porch roof, are iron columns, most of them
marked with the letter P. These are water pipe—and almost certainly were made at Samuel Richards' Weymouth Furnace, which
had large contracts for such pipe with the City of Philadelphia's
water works. It is interesting that even in those days Philadelphia
sent inspectors to check the quality of production. The P marking
was to distinguish pipe made to Philadelphia specifications from
the pipe which was produced for other cities. The birth date of
the mansion—1826—was long proclaimed in plaques above the
downspouts at the front. Sills of the windows are made of iron,
and for some years there was an added north veranda. This was
removed in restoration.

A broad center hall runs the full depth of the building to a door
at the north side, where the handsomely restored portico was long
the carriage entrance. At the left of the center hall, facing north,
are two large and imposing rooms with wide connecting doors,
which can be thrown open to make what probably served as a ball-
room. Little imagination is required to picture gay couples gliding
gracefully through these rooms, in the bright light of a multitude
of candles, while across the hall a spacious dining room offered
refreshments befitting the standards of one of the wealthier men

of his day. Midway along the center hall is the narrow staircase, somewhat Jeffersonian in character and the least attractive feature of the house. It turns sharply at right angles, and a reverse turn at the half-story landing is brightened by a large window, the only window on the east side of the house. In all, there are 13 rooms, the four on the third floor having semicircular windows.

In addition, there is the ground-floor basement, in the "plantation style." Here the housekeeping activities centered. The kitchen was equipped with a great brick hearth, with a small, built-in bake-oven. Nearby is what once was a meat room, and adjoining that a milk room, both dark as night and cool even on a warm summer day. Originally the house was heated by fireplaces, one in each of the larger rooms and all connected to the four great chimneys. Later a hot-air furnace was installed. Since the house was used for the most part in warm weather, the slatted-panel doors upstairs, providing ventilation, were probably of greater importance.

Particularly when the sun is low and shadows are deepening, the ravages of time are less evident at Atsion. There may still be seen through the eastern window the location of the "great garden where fruit and vegetables grew as by magic," to quote a contemporary reporter. The same source tells us that Samuel Richards was "particularly fond of sauerkraut, which he made with his own hands." Leaving the kitchen where such culinary rites were conducted, and passing through a door to the eastward, the explorer comes upon an outside wooden gallery, approached by two sets of steps, with space beneath where horses were tied up to await departing guests. There may have been few "modern conveniences" in that Atsion mansion, but at every turn there is evidence that its occupants lived well.

Scarcely second in importance as a monument of the Richards era is the store. Outwardly this small building looks much like a chapel. It has a belfry, and one legend asserts that it was indeed built for a Roman Catholic chapel but for reasons now obscure never was used for that purpose.[11] Records do not support the legend. The belfry, which has deceived many, was built to house the bell which called workmen to their tasks; a similar bell served a similar purpose above the store at Batsto. This Atsion store was built in 1827, and the old account books show it was in operation in that year—as a company store, of course. Coffee then was 16 cents a pound, butter 15 to 25 cents a pound, eggs 11 cents a dozen, and rye whiskey 37 cents a gallon! Oddly enough, in the modern view, sugar was expensive—10 cents a pound.

This old store often served as the post office, which was moved back to Atsion from "Sooy's Inn" on June 13, 1832. Samuel Richards himself held the title of Postmaster for many years. Until recently the old shelves were in place, and some forgotten boxes of tea and cakes of soap were to be found in odd corners, while outside a gasoline pump testified to the fact that this store survived to modern times. There were even posters bearing rationing regulations of World War II. In 1946, however, the doors finally closed. In one of them a hole had been cut, to let the cat in and out. Nowadays, the old store serves as an office for the agent in charge, locally, for the State of New Jersey.

Atsion's growth under the Richards regime, and particularly its iron output, is suggested by this note in T.F. Gordon's *Gazetteer of the State of New Jersey* for 1834:

> Besides the furnace there are here a forge, grist mill and three saw mills. The furnace makes from 800 to 900 tons of castings and the forge from 150 to 200 tons of bar-iron annually. . . . There are about 100 men employed, and between 6 and 700 persons depending for subsistence upon the works.

From the old Atsion Furnace account books it is possible to piece together a fairly accurate economic picture of the times in general and Atsion life in particular. Furnace tenders in 1825 were paid $25 a month, laborers 75 cents a day, carpenters 78 cents a day. One William Jones was paid $2 for carting a ton of pig iron to Lumberton in 1825, and Jacob Emmons (who also worked at Martha Furnace) supplied a two-horse team for carting wood at $1 for half a day. Also listed is a charge of $4 to Thomas Rubart—for two weeks' board, at $2 a week!

Under Samuel Richards, Atsion appears to have been somewhat less of a feudal barony than was Batsto under his father and his brother, Jesse. However, Samuel accepted responsibility not only for operation of his ironworks, but for the welfare of the community built around them. Some idea of his point of view and breadth of vision may be gained from an occurrence at his Weymouth Furnace when business was at a standstill in 1840. Richards kept his men at work and accumulated a large stock of iron pipe without having found a market for it. His resourcefulness is shown by what followed. Mobile, Alabama, was starting to build a water works. Richards subscribed for stock—on condition that he could pay for it with his surplus pipe. The bargain was made, and his men kept on working.[12]

Connected with the Atsion enterprise in all probability were two "mystery forges" off in the woods to the east. One of these was located by the Atsion River about two and one half miles below Atsion. Here once were locks used by ore boats headed for Batsto. The other was beside the Batsto River not quite two miles above Quaker Bridge. This was a forge called Washington, and was connected with Hampton Furnace, farther upstream. There are banks of slag at each of these locations, while some remains of a dam may still be seen at the site above Quaker Bridge. It is now known that these installations, operated during the Richards era, and Washington Forge [not to be confused with Washington Furnace at Lakewood] was in use as late as 1850.

When Samuel Richards died, on January 4, 1842, Atsion was at the peak of its prosperity. The working force numbered more than 120 men, a Methodist church had been established, and there were some two dozen dwellings. Yet dark days were at hand, and it is doubtful whether even the resourcefulness and dynamism of Samuel Richards himself would have been sufficient to cope with the economic crisis which came with competition from the anthracite furnaces of Pennsylvania. In his will, probated on January 11, 1842, Samuel Richards left to his daughter, Maria Lawrence Richards, "one half part of my Atsion Works Estate and all lands originally purchased therewith by me." This included the Hampton and West Mills tracts. The other "one half part" was left to his son, William Henry Richards. In 1842 these were handsome legacies. A few years later they were something else again.

On June 14, 1849, a little more than seven years after her father's death, Maria L. Richards was married to William Walton Fleming, of Charleston, South Carolina. With the Richards family to sponsor him Fleming soon became a considerable figure in the business world. In 1848 he had launched the W.W. Fleming Cobalt and Nickel Works, located on Cooper River in Camden.

He was one of the organizers of the Camden & Atlantic City Railroad and served on its board of directors for several years. Presumably it was due to Fleming's influence that when the first train, of nine cars, made the run from Camden to Atlantic City on July 1, 1854 (in two hours and a half), the shining new locomotive was named "Atsion."

Fleming and his wife lived at Atsion, in the old mansion, although they maintained a Philadelphia residence at 347 Arch Street. Aided by his brother-in-law, Fleming endeavored to carry on after the bog-iron industry's collapse. Inspired perhaps by the success of paper mills at Harrisville and Pleasant Mills, Fleming erected a paper mill at Atsion, by the site of the old ironworks. Two partners in this venture were Walter Dwight Bell, who had married Mrs. Fleming's sister, Elizabeth, and Albert W. Markley, a Camden banker. This paper mill was in operation a very short time, probably less than two years, and most historians have doubted its existence. There can no longer be any doubt. The mill is plainly marked on the 1860 map of Stone & Pomeroy. That it was erected at least six years earlier is shown in several legal actions, including one for $920 on a trespass and building lien plea by Merrick Murphy against Fleming, Richards, Bell, and Markley, described as co-owners "of the Paper Mill at Atsion." The summons, issued in April, 1855, gives the following description of the mill:

> The said building is a stone mill erected for the manufacture of paper. The main building is two stories high about sixty feet long by fifty feet deep, and attached thereto and making a part thereof are a boiler and bleach house, forty-two feet by thirty-two, a machine house eighty feet by twenty-four, a water wheel and a wheel house, twenty-eight feet by twenty-four.

Atsion's paper mill probably was built in 1852 or 1853. In 1851 Fleming was listed in Kirkbride's New Jersey Business Directory as owner of a "Metallic Works" in Camden, and as a "lumber dealer at Atsion." No paper mill was mentioned. Thus, the time of its erection lies between that date, 1851, and 1854, which is the date of one of the legal summonses mentioning the paper mill. In fact, the phraseology of the summons above suggests that the paper mill already had ceased operation, or, possibly, had not operated at all. The mill is not mentioned in connection with the sale of the Atsion property in 1861.

Fleming's luck was consistently bad. Everything he put his hand to seemed to fail. Atsion had been caught in the wake of the iron industry debacle. His paper mill scheme appears to have been doomed from the start. He lost heavily, along with many others, on the seashore railroad venture. His important position had enabled and encouraged him to borrow extensively. Then came the grim year of 1854, a year of widespread business failures and bank closings. Newspapers of 1854 record many suicides. In the fall of that year Fleming's own financial house of cards collapsed. On September 11, 1854, he made an assignment of his assets "for the protection of creditors." The assignees were his partners, Bell and Markley, his father, Thomas Fleming, and another brother-in-law, Stephen Colwell. Fleming and Colwell, however, refused to serve, after which Bell and Markley became the sole assignees. Roughly half a million dollars was involved, and Fleming's list of creditors filled four solid inches of newspaper type. It was an impressive list, which included the names of many prominent persons and imposing financial institutions.

Desperate, he fled the country, and for a whole year his family had no idea of his whereabouts. He had gone to Belgium, and there, after a reconciliation, his wife rejoined him. They lived the remainder of their lives in Brussels. That Fleming was fundamentally decent, and not a scoundrel, is suggested by the fact that Maria's mother, Samuel's widow, went to Brussels to live with them, and other members of the family made frequent visits. But all of Fleming's numerous properties were lost. His Cobalt and Nickel Works was sold under the hammer and in 1860 passed into the hands of Moro Phillips, who established a chemical works on the site. Fleming lost his wife's share of Atsion and the mansion which had been their home. Gone were various other properties: dwellings in Camden; a house on Walnut Street in Philadelphia; timber tracts in Shamong and Waterford Townships; and even two "Atsion Wharf" sites (Nos. 4 and 5) on the Rancocas in Lumberton, where many of the furnace products had been shipped in the old days.[13] Gone, too, was much of his wife's substantial fortune; and his brother-in-law and ex-partner, William H. Richards, moved to a farm nearby on the old stage road and spent his few remaining years tilling the soil.

Following foreclosure and other legal proceedings, the "Atsion Estate" was offered under the hammer on April 7, 1859. The *West Jerseyman* gives this lush account of the event:

> The sale of the Atsion Estate, at the West Jersey Hotel . . . was the largest public sale of Real Estate which has probably ever been made in this section of New Jersey. The extent and value of the property, the widespread and diversified interests involved by the transactions of its late owner, served to draw together a concourse of Brokers, Bankers, Real Estate Operators, Lawyers, Speculators and Capitalists, more in keeping with the Rotunda of the Exchange than the quiet parlors of a Country hotel. Presently the "frosty pow" of Mr. Thomas, whose head is silvered o'er in the service of the fatal hammer, was observed to rise amid the crowd, and announce that "Atsion", its mansion, its mills, its buildings and broad acres, were positively to be sold, without reserve, to the highest bidder. The terms of payment were stated, the sale to be made subject to a mortgage of seventy-five thousand dollars. An awful pause ensued, during which the good looking company looked around for the brave man who would venture a bid upon it. He turned up in good time, however, with a bid of $5000, when the bidding went on pretty spiritedly, at $1000 a bid, between two gentlemen only, till it reached the sum of $33,000, when the veteran auctioneer took breath, and resumed, with a little professional expatiation, which produced another bid, when the property was knocked down to M. NEWKIRK, Esq. for the sum of $33,500, which, including the mortgage of $75,000, will bring the cost of the whole 28,000 acres to the round sum of $111,500. . . . [The mortgage figure of $75,000 is correct. Probably the first figure should be $36,500.]

For reasons still obscure, this transaction was never completed; no deed is recorded in the name of "M. Newkirk, Esq." Instead, another Atsion auction was advertised for January 3 of the following year. No bidders appeared on that occasion. The sale was adjourned to January 31, 1860, but still the property went begging. Not until April 13, 1861, was a sale concluded.[14] The purchaser was Jarvis Mason, of Philadelphia. The price was $66,000. The Camden *Democrat* observed "it is considered a great bargain."

Since the mortgage alone had been $75,000, it certainly was a bargain. Moreover, Jarvis Mason was one of the few, after Samuel Richards, to make money out of Atsion. A year and a half later, Mason sold Atsion to Colonel William C. Patterson, of Philadelphia. The date was July 11, 1862. The price was $82,500.[15]

Colonel Patterson seems to have had a few plans of his own for Atsion. He organized a land development company, renamed the town "Fruitland," and offered lots and farms to the public. Patterson is credited by Woodward and Hageman, in their *History of Burlington and Mercer Counties*, with rebuilding the church and establishing the present burial ground. A few tracts were sold, most of them off in the woods, but soon Patterson, like so many before him, met with financial difficulties. There followed the familiar resort to an assignment for the protection of creditors, and Atsion once again became a grave of enterprise and fond hopes.

As late as 1885 an observer, quoted in Harshberger's *Vegetation of the New Jersey Pine Barrens*, wrote "From Manchester [Lakehurst] southward to the Mullica River is one of the wildest, most desolate portions of the State." It was through this wilderness, however, that the stimulus for Atsion's next boom was to come. Those were exciting days of railroad pioneering and expansion. In 1856 William and John Torrey, who had large tracts of land in Ocean County, launched their Raritan and Delaware Bay Railroad. They planned a line roughly bisecting New Jersey, connected with New York in the North by ferry from Port Monmouth and with the South by boats which would ply the waters of Delaware Bay.[16]

The Torreys had frequent financial difficulties, and not until 1860 or 1861 were they able to cut through the thick pinelands to carry their tracks as far as Atsion, which became their temporary southern terminal. Next, a connecting line was run through the woods to Atco, on the Camden & Atlantic Railroad (later to be taken over by the Pennsylvania Railroad). The purpose of this spur was to provide direct service from New York to Philadelphia. Here, however, the Torreys were in competition with the Camden & Amboy Railroad, one of the most ironclad monopolies in American history. Its President, Commodore Robert F. Stockton, boasted: "I carry the state in my breeches pocket, and I mean to keep it there."

The Torreys began operating trains from New York to Camden, via Atsion, in 1862. They had the encouragement of the United States Secretary of War, because the new route was useful for Civil War transport of soldiers and military equipment, at a time when the Camden & Amboy was overburdened. A long legal battle followed. The force of Commodore Stockton's boast was demonstrated when the New Jersey courts obediently granted an injunction blocking the Torreys' operations, and did so even though the United States Congress passed legislation making the Raritan & Delaware Bay Railroad a post and military road, thus officially supporting its right to compete for the Philadelphia trade. By 1867 the end of wartime prosperity, along with other woes, brought the Torreys to bankruptcy. It was when their line was reorganized as the New Jersey Southern, now affiliated with the Jersey Central lines, that there was talk of a new boom in Atsion and other pineland places. Atsion, particularly, seemed a likely spot for industrial development, since it had rail links to both New York and Philadelphia.

Colonel Patterson had assigned his assets and those of the Fruitland Improvement Company to George W. Dallas. At the usual Masters Sale, Joshua L. Howell officiating, the Atsion Estate was sold on May 10, 1871, to Maurice Raleigh, a wealthy Philadelphian, who had been born in Ireland, of humble parents, and had come to the United States while a small boy. The price for Atsion this time was $48,200. Raleigh was fascinated, as others had been, by the possibilities of the Jersey pinelands, and he proceeded to move with much the same resolute vigor that had characterized Samuel Richards. Raleigh rebuilt the great mill into the structure it is today, and fitted it up as a cotton factory. It is a cavernous building, stoutly buttressed with solid, foot-square beams. Using cotton brought from the South by rail, this plant soon was turning out five hundred pounds of yarn a week, with some 170 persons employed at the mill alone. The same convenient rail transport carried the finished products away.

Raleigh's mill appears to have been profitable, and to have operated at capacity during most of his lifetime. Raleigh also built a carpenter shop and a blacksmith shop, and himself became operator of the general store. In May of 1875 that store was "robbed of upwards of $300 worth of boots, shoes, hats, shawls, jewelry, cutlery and other goods." News accounts stated that "The thieves carried away their booty on a hand car." [17]

During Raleigh's era a school was in operation—Public School No. 94—and teachers there were paid $30 a month which seems to have been the prevailing wage for rural instructors at that time.[18] The school building still stands. The little chapel had become the "Free Union Church," used by all denominations. As for industrial Atsion, an idea of that is given by the *New Jersey Business Directory* for 1876:

Brown, B. B.	Postmaster. Flour and Feed Store
Brown, Miss E. E.	Millinery and notions
Hagerthy, G. W.	General store
Hyland, Thomas	Carpet weaver
Raleigh, M. & Son	General store
Thompson, E. T.	General store
Wells, Jas.	Wheelwright

Note that three general stores are listed for 1876. There is none today

Another of the many abortive experiments involving Atsion was a project, started in 1881, to establish a "colored colony" there. The Camden *Daily Post* of April 15, 1881, reported: "A colored delegation, headed by ex-Congressman Smalls and a Bishop of North Carolina, visited this place again . . . with a view to purchase of the entire estate for the purpose of establishing a colored colony there. The East, South and West are represented in the delegation."

Five days later the same paper noted that Raleigh had "offered to sell the property for $1,300,000, and the first payment, $150,000, is to be made August 1, 1881. A similar amount is to be paid April 1, 1882, and $250,000 April 1, 1883. The balance is to be placed on mortgage for six years." Needless to say, this ambitious project never was realized, and few people seem to have heard of it.

Maurice Raleigh died on January 10, 1882. He left Atsion as an asset rather than a liability to his estate, which also included valuable tracts at Waterford. As a result of various purchases, and exclusive of some land he had sold to Joseph Wharton in 1873, there were more than 30,000 acres in the Atsion Estate at the time of his death. Thus, his heirs had every encouragement to carry on, with one John O'Dea remaining as manager of the cotton mill. Unfortunately, as so often is the case with heirs, they lacked that sure touch which Raleigh had possessed. Before long the cotton mill was running at a fourth of capacity, and later the same year it closed, forever. Then, like others before them, these heirs turned to a new field in the hope of recouping what they had lost.

Raleigh seems to have been responsible for having the name of the town changed back from "Fruitland" to "Atsion," on August 1, 1871. His heirs took steps to change it again, to "Raleigh." On February 10, 1885, the Raleigh Land and Improvement Company was formed, when Raleigh's five children joined with a syndicate including several New York men and Raleigh's son-in-law, James G. Fitzpatrick. They moved to lay out the tract in building lots and small farms.[19] On a "Map of the Town of Raleigh, Formerly Atsion," by Frank Earl, surveyor, the town is described as follows:

Atsion Junction, 27 miles from Philadelphia on N.J. Southern Division of N.J. Central R.R. (operated by Reading R.R.). Trains leave New York from foot of Liberty St. (N.J. Central R.R.) and leave Philadelphia from Pier 8 South Wharves, Walnut St., by Philadelphia and Atlantic City R.R. (operated by Reading R.R.). Town lots, general size 75 feet front and not less than 150 feet deep, and farm lands adjoining the town for sale. Fertile farm lands at $25 per acre and on five years' time to actual settlers. No liquors. Excellent soil, good railroad communications with low rates, freight leaving Raleigh at 8 p.m. reaching New York shortly after midnight. Ground reserved for Schools, Churches, Parks &c. A beautiful lake, 1½ miles long, with wooded banks and park adjoining, with boating, fishing, bathing &c. Very fine sites for locations of winter or summer health resorts. Pure air, good water, and absolute freedom from Malaria. Good accommodation for visitors.

The last sentence suggests that the Raleigh Company planned to use, perhaps actually did use, the old Richards Mansion as a hotel. The map, incidentally, shows two railroad stations, only a few blocks apart, one marked "Raleigh," the other "Atsion." On the plan Atsion Lake is renamed "Lake Roland." The Tuckerton trail becomes "North Lake Avenue;" other streets are named "Sewell,"

"Haines," and, in a gesture to the past, "Richards Avenue." A
flour mill is shown; nearby there are a building marked "Factory,"
a "Gas Works," and a "Blacksmith Shop." The old church is
noted as "Union Church Property." The course of a canal from the
lake, to power the cotton mill, is clearly indicated, and its route
may be traced today.

Some idea of Atsion at that time may be gained from the
Burlington County Directory for 1887, which notes:

> ATSION: Village of ab. 200 inhabitants. Small stream furnishes
> power to run a grist mill and cotton factory.

Cotton, Thomas	Justice of the Peace
Lloyd, B. F.	Station Agent
Lloyd, M. F.	General store
Miller, Theodore	Constable
Raleigh Land & Imp. Co.	J. Fitzpatrick, Pres.
Stiles, E. E.	General store
Stiles, Thomas B.	Postmaster, Telegraph, W. Union

The efforts of President Fitzpatrick were not very successful.
Some few plots were sold, but so short-lived was the "Raleigh"
boom that few today know that it ever took place. As the name was
not legally changed, the chief surviving evidence is the map of the
development and some privately owned properties on the north
side of the street above the lake, which, happily, is still Atsion Lake
and not "Lake Roland," just as the original town name endures,
with "Fruitland" and "Raleigh" all but forgotten.

After the brief real estate boom, if it can be called that, finally
fizzled out, it was not long before Joseph Wharton (in 1892)
acquired the Atsion tract—excepting the few properties sold above
the lake—and added it to his vast holdings in the Pine Barrens.

Today Atsion remains a quiet community. The great mill still
stands, in efficient use for processing cranberries. Aside from the
modern lake cottages there are, below the Richards mansion and
store, some half-dozen houses built of cedar. Two of these are
double houses and probably date from Raleigh's regime. Of these
six venerable dwellings, no less than five are equipped with tele-
vision aerials, the most modern touch in Atsion. A log cabin of
mysterious vintage stands sturdily near what once was the main
millrace. The old church is now an Episcopal mission, and back

on the road to Quaker Bridge—and Tuckerton—are four more houses; one of these is said to have been the schoolhouse, and its appearance supports that tradition. Across from them is the one-time manager's house. And that is about all. The railroad tracks over which soldiers once rode to Atco and on to Philadelphia have long since been ripped up, but the path they followed is clear. Trains run through on the main Jersey Central tracks, one train each way each day, but they do not carry passengers any more. Even the once-glamorous "Blue Comet" seashore-special no longer operates. Those who look for Atsion Station may still find it—but not in Atsion. It can be seen, several miles away, in the middle of a farm, where it is used as a pickers' shack.

Much like time itself, however, the Atsayunk flows serenely on, giving majesty to the great lake, spilling over the dam into foaming eddies, seeking its familiar course, as it has over the centuries— through the pinelands, into the Mullica, and on to the sea.

5

QUAKER BRIDGE TO
WASHINGTON

For those who nourish a secret eagerness to escape from "progress"
—those whom thoughts of Thoreau tempt to try living in the
woods, and Whitman's strophes urge toward "the open road"—
a trip through Quaker Bridge is a major experience. The old stagecoach road to the seacoast is little changed from what it was a
century and more ago. Some would call the region through which
it passes "desolate"; a better word would be "subtle"; nature's
mood, throughout the Pine Barrens, is keyed low, relaxed, such as
to "invite the soul."

There is only one crossroads in Atsion. Pointing east from
U.S. 206 is a directional arrow which reads: "QUAKER BRIDGE
4." This sign is deceptive, for anyone expecting to find, at Quaker
Bridge, a town, a hamlet—or even a house—will receive the surprise of his life. That way, however, lies the old road, which takes
a sharp dip from the concrete of U.S. 206, winds by the silent old
Richards mansion, passes several weather-beaten houses and the
little white mission of St. Paul's. Then comes a hump over the
tracks of the Jersey Central Railroad. Here, since there is but one
train daily each way, it is reasonably safe to pause for a glance at
the impressive swath which those tracks cut through the Pine
Barrens. To the north, the eye can follow the ribbons of rail until
they finally converge in a forest-shrouded infinity.

The road dips again and then cuts directly into the woods, runs straight for a bit but soon curves gently to the left. For most of the journey, the road is a single sand-track, but shortly the traveller comes upon a double-lane turnout, which is perhaps a thousand feet long. This "siding" he instinctively dates from the days when not only stagecoaches, but trucking teams bearing cannon balls, water pipe, pig iron, charcoal, and lumber, required a convenient place to pass. At intervals, now, the sandy tracks cut deep, while in some sections they are almost level with the forest table which stretches endlessly on either side. Often—too often—the road passes through areas badly and rather recently burned. Gaunt, denuded tree trunks reach for the skies amid blackened stumps and devastated earth, while here and there shoots of pitch pine renew the old struggle for forest survival. Soon all is green again, usually with a wealth of laurel and other shrubs softening the shadows of oak and pine. Occasionally aisles carpeted with moss veer off to the right or left, where loggers of long ago made tracks scarcely visible now.

Many of the turnings in the trail seem to have no reason for being. But the resulting vistas are pleasing to the eye, and it is at such places that deer are likely to dart across the path, usually well ahead of the car, but sometimes frighteningly close. The modern automobile moves smoothly and almost silently on the firmer sandy roads. It is like riding on velvet, so that with car windows down, the traveller frequently may hear deer crashing through the underbrush, even when they remain unseen. On other occasions, they show up in force, and on one banner day no less than seventeen were seen bounding across this very trail.

After a little turn past a cedar thicket, not quite two miles beyond Atsion, a bit of civilization punctuates the wilderness. It is a cranberry bog, not a large one as bogs go, but well fed by a briskly flowing stream, which is noted on most maps but not named on any seen by the author thus far. There is a small dam, water spilling from it under the narrow bridge which carries the road. Flowing on through tangled cedars, this little brook soon joins the Atsayunk—now officially called the Mullica River—which is not far to the right at this point. From here forward, however, road and river diverge, the Mullica turning almost directly south, while the trail bends well to the southeast.

Over the next mile and a half the terrain is more or less level.

and yet the trail winds constantly. Thick underbrush, pressing close to the road, beats a sharp tattoo on the sides of the car as it takes the curves, scarcely improving the paint job. Frequently, a dead tree has cracked apart, scattering its rotted limbs far and wide and leaving the larger branches still striking crazy angles against the living trees nearby. If the sun is bright—and it is best to make this trip when it is—the taller trees cast majestic shadows on the white sand of the trail. There are places where the ruts may be deep enough to cause concern if the car is particularly low-slung. Generally, however, in dry weather at least, the sand is quite firm over this first part of the journey, and the center road clearance is adequate.

Just beyond the cranberry bog, an inviting road leads to the right, with a small clearing at the junction. This road goes to Batsto, is shown on maps dating from 1778, and undoubtedly was in use long before that. Over it must have been carried the pig iron which in the early days of Atsion Furnace was sent there from Batsto for conversion into bar iron. Almost certainly it was one of the routes by which munitions were conveyed to Washington's armies in Pennsylvania from the Batsto works. The main road, which bears left at this junction, follows a somewhat straighter course for almost two miles and is then joined by another—or its remnants—forking in from the left. This is the largely overgrown and forgotten end of what once was the Shamong trail; its travelled stretches, miles to the west, today are known as Stokes Road. The junction is easy to miss, even travelling slowly, as one should in these parts. Then, only a few hundred feet or so ahead, around a bend to the left, comes the very heart of New Jersey's "Wharton Estate"—Quaker Bridge!

There is little here but history, and the singular beauty of a lonesome river. It is enough. Quaker Bridge had its chance to become a "big town"—or at least a town—back in 1836. Those were days when railroads were seeking new fields to conquer and plans were afoot for a through route from Philadelphia to the beckoning resorts of the Jersey seashore. On March 10 of that year a charter was granted to the Camden and Egg Harbor Railroad Company, which boasted an authorized capital of $200,000, to build a railroad along a route from Camden to Quaker Bridge, thence to McCartyville (later Harrisville), and on to Tuckerton.

Among the incorporators were: Jesse Richards, lord of the manor at Batsto; Ebenezer Tucker, of Tuckerton; William McCarty, who had only recently started his paper mill at McCartyville; Samuel B. Finch; and Timothy Pharo.[1] Luckily or unluckily for Quaker Bridge, this project fell through. The line was never started, and a railroad to Atlantic City was constructed instead. Thus today there is no railroad at Quaker Bridge, no town, not even a house. It is a place name without a place. There is, of course, the bridge, and of that there is a bit to tell.

The ground dips as it approaches the white bridge, after the turn past old Stokes Road. The bridge is framed by a picturesque arch of trees. Patches of swamp lie on either side, and remnants of old stone abutments suggest earlier adjacent construction. The present bridge, indeed, is comparatively new, having been built in 1930. It crosses the Batsto River, which at this point is a sizeable stream about thirty feet wide—cedar water, of course, the color of strong tea and soft to the touch. It is a charming and restful spot. Upstream the view is of a smooth-flowing river, banked by lush and stately cedars, with a mysterious bend which all but commands exploration by canoe. Not too far upstream are the remains of an old dam, which once belonged to the already-mentioned Washington Forge. The dam itself is mentioned in yellowed old deeds as far back as 1803. Downstream the way is blocked by fallen pines and cedars in substantial number. Some years ago canoes are known to have taken off from this point, but without hatchet work that would not be too easy today.

This stream caused the events which gave Quaker Bridge its name, back in 1772. It had been a custom of Quakers to combine devotion and relaxation by making annual trips to the meeting at Tuckerton. Most of them came from Burlington, Mount Holly, Moorestown, and Upper Evesham, now Medford, and the chief hazard of their journey was the crossing of the Batsto River. For some years they had been accustomed to swim their horses over the stream, but on several occasions the riders were drowned. After one such tragedy a meeting was held on the riverbank in 1772, attended by Quakers from Tuckerton as well as the other communities mentioned. It was proposed to build a bridge, and after the meeting of that year a band of those attending gathered at the ford, cut down a sufficiency of the nearby cedars, and set up a "substantial bridge." Thus, having been built by Quakers, it was named Quaker Bridge.

Thirty-six years later (although the date once was in dispute), Quaker Bridge staked its second claim to fame. The curly fern, *schizaea pusilla*, was discovered there, an event which brought botanists from far and wide, even from Europe, to this remote spot in the pinelands. A label accompanying a specimen in the collection of the Torrey Botanical Club has this to say about it:

> First discovered by Dr. C. W. Eddy, near Quaker Bridge in the pine barrens of New Jersey, about 30 miles from Philadelphia. Dr. E. was in company with J. LeConte, Pursh and C. Whitlow and though he and Mr. LeConte found all the specimens, Pursh has claimed the honor of the discovery himself.

Pursh generally has been credited with the discovery, perhaps properly since he was first to place the fern in its right genus. In their book, *The Ferns of New Jersey*, M.A. Chrysler and J.L. Edwards give the date as "about 1808" and note that this "least fern-like of our ferns is the most famous New Jersey plant." They also quote the botany manual of Amos Eaton to the effect that "A party of botanists, consisting of LeConte, Eddy, Pursh and Whitlow found three specimens only of the species, in 1805, all of which have been lost." They comment that "It thus appears uncertain whether 1805 or 1808 was the actual date of discovery." This rare plant since has been found near Toms River, at Pleasant Mills, and some 27 other scattered spots in the Pine Barrens.

If Quaker Bridge became a favorite spot for naturalists, it also seems to have been a mecca for pleasure-seekers, if half the legends about the place can be credited. From early days, before 1810 at least, Quaker Bridge had a hotel which served as a town-meeting place and election headquarters, besides being one of the principal stopping places of the stagecoaches on their way to and from Tuckerton. Even today folks as far away as Lower Bank remember their elders telling of weddings, oyster suppers, and other social functions which were held there.

It is known that in 1818 the host of the Quaker Bridge hotel was named Thompson. One of the few written records of a visit there was made by the botanist John Torrey in a letter to Zaccheus Collins. This letter tells of a trip he made with William Cooper through the Pine Barrens and is so interesting that some extracts are given herewith:

N.Y. July 9th, 1818

We arrived at South Amboy one week after we left Philadelphia, and although our journey was rather an arduous one, we think ourselves well rewarded for the privations we endured. The principal difficulty was in keeping the right road. Hundreds of these little roads cross each other in every direction like a labyrinth, so that it is next to a miracle if you hit the right one. We remained two days at Thompson's Tavern (at Quaker Bridge) where we were very well entertained. About this time we found a considerable number of plants which were new to us. . . . What pleased us more than any plants we found was the Schizaea. Cooper found the first specimen. It is a singular little plant, and I first doubted whether Pursh had referred it to the right genus, but subsequent examination has convinced me that he is right. . . .

After we left Quaker Bridge we fared pretty hard. Some places called taverns that we put up at were not fit for an Arab. At a place called the Ten-Mile Hollow, or Hell Hollow, we expected to sleep in the woods, for it was with difficulty that we persuaded them to take us in. This was the most miserable place we ever saw; they were too poor to use candles. No butter, sugar, etc. A little sour stuff which I believe they called rye bread, but which was half sawdust, and a little warm water and molasses, were all we had for breakfast. For supper I could not see what we had, for we ate in the dark. From this place until we reached Monmouth we found scarcely a single plant in flower. . . .

The Quaker Bridge tavern was still standing in 1849, when it was shown on maps and listed in Gordon's *Gazetteer*. Also shown in that year was the house of "W. Richards" across the road from the hotel, at a location where it is still possible to trace the vague outlines of an old cellar hole. Just when the tavern burned down—its probable fate—is as difficult to discover as the date of its erection. Certainly there is little now to suggest human habitation, at any time.

There is a "labyrinth of roads" much like that Torrey found a century and a quarter ago; so, too, in season, is an abundance of the golden *hudsonia* which he also remarked. The fact that

Quaker Bridge is a junction of no less than four trails, all of which may once have been important, perhaps explains why in wagon wheel days the tavern was as busy as legends tell us. The first junction road swings sharply left, to run more or less northwest above and along the bank of the Batsto River. It winds eventually to what once was Hampton Furnace and now is a major cranberry development, and on the way passes a clearing where the mysterious forge may have stood. For a motorist, it is a road to be "handled with care." The second trail on the left leads to High Crossing, on the Jersey Central, which is not far from the spot where the Mexican airman, Emilio Carranza, perished on his famous good-will flight in 1928. The cellar hole and a few bits of brick and fieldstone are in the angle formed by these trails. What seems the most likely location of the old hotel is the substantial clearing at the right of the main trail, which heads more or less due east, but which looks so much like all the other branch trails that many travellers have been lost in these parts since Torrey's visit. On a fine spring day this clearing has seemed to be almost solidly covered by the *hudsonia* or, as some call it, "golden bell." The fourth road turns sharply to the right a few hundred feet beyond the bridge. This, too, winds along the Batsto River, but soon turns and re-connects with the main trail. Some years ago a hunter's shack stood at this spot, but such is the capacity of the Pine Barrens for obliterating the handiwork of man that no trace of it remains. Aside from the bridge itself, the sole evidence of civilization in Quaker Bridge today is a concrete surveyor's monument.

If historical information on Quaker Bridge seems scarce, there are plenty of legends. "Old-timers" have told tall tales of the famous outlaw, Joe Mulliner, waylaying stagecoaches near Quaker Bridge. One such tale has it that Mulliner strode into the tavern there, shouted to the fiddler, halted the dance, and demanded a waltz with the prettiest girl in the place. This seems to have been a custom with Mulliner. On the occasion in question, however, when he ordered all the men out of the room, one fellow worked up enough nerve not only to defy the bandit, but to clout him in the face. Mulliner, delighted, grasped the boy's hand, laughed heartily, and dashed off without further ado. There is one difficulty with this captivating yarn. Mulliner was hanged in Burlington in 1781. There is no evidence that the stages were using the Quaker Bridge route at that early date, and there is nothing to indicate that the Quaker Bridge hotel was in existence in Joe Mulliner's time.

Another legend, and one of the best, concerns the white stag which was said to haunt Quaker Bridge, and whose appearance, in a departure from similar legends elsewhere, foretold not death, but danger, and thus was a warning rather than a sign of inevitable doom. Cornelius Weygandt in *Down Jersey* has told briefly of a "White Stag of Shamong" which was observed in the vicinity of Indian Mills, once called Shamong. In 1953 a white stag was reported seen near Chatsworth, a town which at one time also bore the name of Shamong. The white stag of Quaker Bridge, however, has a long history; it was seen chiefly by stage drivers, always in the vicinity of the bridge itself, and usually on the west side.

The following tale is set against the dark skies of a late summer day, when a great storm had banished a morning of sunshine and whirled torrents upon the countryside. Those who know the Barrens region know how its streams can rise to startling levels, how winds can fling whole trees into the waters, and how the waters—high-crested, swift, timber-laden—can sweep all before them. So it was on this night in the last century. Heading east, the stage was "so near" to the haven of the Quaker Bridge hotel, and yet "so far" from it. The anxious driver strove to hurry his horses in the full fury of the elements, while the little band of passengers huddled inside a coach scarcely proof against such weather.

Suddenly a light shone ahead, a light dim in that storm, but bright enough to tell the happy coachman that it came from Quaker Bridge. Then an extraordinary thing happened! Out of nowhere came a white stag—which halted in the very center of the road. Even in the darkness, the legend tells, there could be no doubt what it was. The coachman reined in his horses. At first he was angry, then puzzled, then alarmed. He could still see the light, but as he left his seat to investigate—the stag vanished. Perhaps the coachman was superstitious. He may have been merely cautious. In any case, despite the storm, he tramped down the soggy road to see what lay ahead. He was not long in finding his answer.

The bridge had been torn from its foundations, and had he dashed forward impetuously toward the beckoning light—beyond the stream—the whole equipage would have plunged into the river.

A night under the stars at Quaker Bridge might conjure up still more of its past, but there is the trail to follow. A swift start is needed to carry the car through a hundred feet of really deep sand. Beyond, however, the trail hardens, much as before. It rises over a slight hill, dips a bit, then winds on almost directly eastward, while the Batsto River flows, roughly parallel with the Atsayunk, to the south and southeast. Here indeed is the heart of the Pine Barrens. On a calm day the stillness is amazing, broken only by an occasional calling of birds or by a vagrant breeze flipping the green branches far above. The traveller may have gone six miles without encountering another human being, and may go almost as far again before doing so. True, he is but six miles from a concrete highway to the seashore, but he might as well be in the Canadian wilds as far as any interference from civilization is concerned. Road signs are nonexistent, and most maps offer scant help. Here is peace, and contentment. Some, of course, will say the country is dull and dreary, barren in fact as well as name, and what can anyone possibly see in such desolation? People of that sort had best stick to the concrete "trails." This one is not for them.

About two and a half miles beyond Quaker Bridge there is a significant bend in the road. With this turn, there comes a sense of awe, almost foreboding. Even on a brilliant day, barely past noon with the sun high and strong, there is a sense of nightfall, a feeling as if the road were headed for some strange chasm. The reason soon becomes apparent. This is the beginning of Penn Swamp—a thick and vast cedar swamp. The road here is under water part of the year, and while fording the swamp is supposed to be safe enough, the quick change in atmosphere and terrain is enough to raise doubts, even in the hearts of the venturesome. With the nearest semblance of human habitation now over seven miles away, caution dictates exploring the road ahead on foot, rather than risk running the car into some mushy cul-de-sac where a turn would be impossible and retreat difficult. In any case, only on foot can the explorer really observe and appreciate this shadowy wonderland. Although a high wind may be blowing outside, in here the air is deathly still. The road is a corridor between lofty cedars, innumerable and thickly grown in a tangle which seems quite impenetrable. A strange cry overhead suggests some odd and unfamiliar bird; it is caused by the towering trunks of dead cedars, abrading each other in the wind. Creaking and moaning, they sway in dizzy dignity. This is no place to linger at night!

Through most of Penn Swamp, which extends for nearly a quarter of a mile, the sun is quite shut out. Only at the edges does it break through here and there to light up a soggy thicket. Along each side of the road pools of cedar water extend into the wilderness as far as the eye can see; apparently they are drained by sluices under the roadway. The road itself, while adequate, has a spongy feel underfoot, like foam rubber. Anyone travelling here, in either darkness or daylight, will soon pay a heavy penalty for straying from that straight and narrow path, for a deviation of merely two or three feet, right or left, would spell disaster.

The lush, exotic vegetation lends deeper drama and excitement to this scene. There is a sense of closeness to elemental forces, to processes of nature which, over the ages, have worked chemical transformations to produce in these dank waters iron and wood with unusual qualities. The iron, once fabricated, is reluctant to rust. The wood, fashioned for use, is impervious to the elements for long periods of time. Ancient tree trunks buried in just such cedar bogs were dredged up, found untouched by rot, and used to build PT boats during World War II. This wood is said to be resonant and excellent material for pipe organs. Here, too, on all sides are beds of sphagnum moss, floating islands of green extending far into the thickets. Gathering sphagnum moss is still a "major industry" in the Pine Barrens. Other mosses are numerous, the common tree moss literally coating whole stumps, so that their most jagged outlines appear soft and delicate. Bluish mosses are quite plentiful in the darker places; and a pink moss of fragile beauty may be found occasionally. Entwined in the cedars is the swamp magnolia, sometimes called sweetbay; its fragrance at blossom time, in June, is legendary. There are other shrubs in abundance; it is easy to understand why a swamp of white cedar— *Chamaecyparis thyoides*—affords a field day to the botanist.

The wandering, sluggish waters here are known collectively as a stream, sometimes called Goodwater Run and shown on other maps as Penn Swamp Branch. The "branch" is explained by the fact that about a mile and a half to the south both stream and swamp merge with the Batsto River. Only the skilled canoeist, however, is familiar with that junction in the jungle.

Well beyond the swamp comes a fork in the road, the two trails being equally inviting in their quiet charm. The right fork winds toward Batsto. The left fork heads toward "The Mount," and for nearly a mile the trail cuts straight through the woods. Then comes a rise which affords a "view," but what a "view"! Most of the trees here have been badly burned, and the growth is scanty, which suggests that the latest fire was but one of many over the years.

"The Mount" is another place name without a place. It offers nothing but scattered evidence of the hotel which stood here in the staging days. No records are available to show when it was built, but its earliest appearance on a map, so far as the writer knows, was in 1849, on the Burlington County map of Otley and Whiteford. There its owner is identified as "J. Crammer," in all probability one of the sizeable family in those parts who once claimed descent from the famous Archbishop Cranmer—various branches spelling their names "Cranmer," "Crammer," and "Cramer." This tavern of "J. Crammer"—the J. stood for Jonathan—was certainly in operation in 1846, and probably earlier. In a letter which the author has seen, petitioning for a change in the Tuckerton mail stage route, it was proposed in 1846 to put a post office at "The Mount" and to appoint "Jonathan Cramer [sic] himself for postmaster."

Remote as it seems now, "The Mount" must have been, like Quaker Bridge, a busy place in the old days, until the stages finally ceased to run. That time came in the 1870's, after the railroad had been cut through to Tuckerton. The last proprietor of the hotel appears to have been Shreve Wills, and in his day it was also a polling place for Washington Township, to which folks from miles around went to cast their ballots.

From "The Mount" to Washington, the next ghost village on the trail, is a short trip measured in terms of distance—about two miles. In terms of time, however, it can be a long journey indeed. On four occasions, the author was balked by boggy stretches in the road ahead. The ruts here were usually covered with water, so that their depth was problematical. On each side of the road ran rivulets, often backed by swampland, so that to turn aside was unthinkable. Given the right weather, however, the trip can be made.

A major crossroad warns of the nearness of Washington. This is the "Batsto-Jenkins Turnpike," as one native of the region has called it. While scarcely a boulevard, it does have two lanes, is firmly gravelled, and by all odds is the best route to Washington from any direction. About five hundred feet beyond this "turnpike," a leaf-bowered trail comes in from the left, and there, at that lonely junction, is Washington itself.

Unlike Quaker Bridge and "The Mount," Washington holds more than ghosts and memories. Not a single structure is standing, but there are ruins and other fascinating evidences of what maps show to have been a considerable village. The first sight to engage the eye is an imposing remnant of what once was a sizeable building, 148 feet long over all. This ruin lies on the right as the traveller enters the "town." The center section consists of a deep cellar, with thick foundations of Jersey sandstone, rough-coated with mortar. This cellar is divided by a great wall, also of sandstone, which rises about 12 feet above ground level, and which originally supported what must have been a substantial roof.

It would be in the romantic tradition to identify this building as a famous tavern, established at Washington before the Revolutionary War, a tavern the prosperity and fame of which endured for nearly a century of usefulness as a major stop for the Tuckerton stages. Gazing at this brown pile of masonry standing out sharply against the forest background, it is not difficult to picture coaches turning in, the colorfully uniformed drivers alighting, the guests entering a warmly lit and inviting hostelry. The facts rudely dispel this pleasant vision. Horses and cattle, not human beings as heretofore supposed, inhabited the stone structure. Leeson Small, veteran caretaker for the Wharton Estate, which included Washington as well as forest and towns for miles around, says that for years there was a cattle-breeding farm at Washington, and that the ruins are those of a combined barn and pit silo, in which the corn was held down by sandbags. Small was among those present when the barn was first threatened by a forest fire, and, again, around 1913, when a second fire destroyed the place. Small's sister-in-law was born in Washington and relates that when she lived there the "town" consisted of this barn-silo, two other barns, and one six-room house.

Originally the place was called "Sooy's," after Nicholas Sooy who founded the real "Washington Tavern," long believed to have served as a major recruiting station for the American army in Revolutionary days. Legend tells that it became a tavern at just about the time the Liberty Bell was proclaiming American independence. In the fashion of those days, the story goes, it boasted a brightly painted sign, bearing a rough portrait of George Washington, encircled by a wreath of laurel, and the inscription: "Our Country Must Be Free." [2]

Frequent visitors to the Washington Tavern were the recruiting sergeants during the War of 1812. Here at evening, they are supposed to have sought out charcoal burners, laborers, teamsters, lumbermen, and ironworkers from nearby Batsto and Atsion Furnaces. A recruiting sergeant would stride into the bar and invite all hands to have a drink. Then he would start a sales talk on the glories of a military career, and wind up by calling for volunteers. It was natural that Washington Tavern became the headquarters for many regional committees, served as a rendezvous for the local militia on "training days," and further was used as a polling place.

When Nicholas Sooy died, in 1822, and was laid to rest in the historic little graveyard beside the Pleasant Mills Church, his son, Paul Sears Sooy, took charge of the Washington Tavern and soon was as renowned a host as his father had been. Until 1820 the stage trip to Tuckerton required two days and Washington Tavern was the principal overnight stop. Even after 1820, when the stages ran through in a day, the drivers usually paused here to change their horses and permit shaken passengers to refresh themselves and stretch their cramped legs. Here, also, a post office was located for some years, after being moved from Atsion in 1815.

The name "Sooy's" seems to have been changed to "Washington" in the 1830's. A map of 1834 shows "Sooy's," and an 1839 map shows "Washington." An 1849 map notes, besides the tavern, two houses, one belonging to "W.H. Sooy" (brother of Paul Sears Sooy). The same house and, of course, the tavern appear on the 1858 map of Kuhn & Janney, as does "Public School 3001."

Difficult as it is to believe of today's woodland, as late as 1882 this region was described in a gazetteer as a "farming district."

What is to be found at Washington now—other than the ruin already described? In front of that ruin, and to the right, are ground-level foundations of what probably was a house, 38 feet long and 16 feet in depth. In front of the ruin, and to the left, about twenty feet from the road, is a small well with bricked walls, still excellently preserved. All other discernible remains are on the opposite—the north—side of the Tuckerton Road. Almost directly across the road from the ruin are two cellar holes. One measures about 15 feet by 30 feet (this is where the school was shown on the 1858 map), and the other roughly 18 feet by 30 feet (some bricks remain here). To the rear are vestiges of an orchard and also a clearing, which suggests that a barn may once have stood there. Some 750 feet east along the Tuckerton road comes another cellar, again on the north side. It is approximately fifty feet off the road and probably was another house, quite possibly that of "W.H. Sooy."

The location of the famous tavern was for a long time a mystery. Its cellar is to be found precisely where some old maps show it to have been: three-tenths of a mile east of the ruin which was not a tavern, at a crossroads where another trail—once another stage road—led from the Eagle Tavern, above Speedwell, almost directly south to what once was Upper Bank and what is Green Bank today. At this junction, almost overgrown, are the remains of the old basement and wine cellar. This identification is not based on guesswork, hearsay, or legend, but upon the old deeds, which describe the "place known by the name Washington Tavern in the Township of Washington . . . on the road leading from Tuckerton to Philadelphia by the Mail Stage Route" and locate one corner of the property as the "middle of the road from Washington Tavern to the Bank [Green Bank]." [3] Bits of the brick and sandstone foundations are still visible here and there, and fragments of brick are scattered under the leaves in the pit. A glance at the trails converging on this spot from four directions invites suggestions of an exciting and busy past. Here, over a century ago, there was much quaffing of ale, clopping of horses' hoofs, and conversation of travellers. They had come to enjoy the hospitality of Nicholas and Paul Sears Sooy,[4] son and grandson of Yose Sooy who emigrated from Holland before America was a nation and whose descendants have become so numerous that they hold large annual reunions in one or another of the towns for which the name "Sooy" has historic associations. They are not likely to hold a reunion at Washington. True, there is a sign at this crossroads even today. It is not a wooden sign, bearing Washington's portrait, but one of metal, urging the utmost care in preventing forest fires. Perhaps the old tavern might be standing yet, if more people had exercised such care in years gone by.

6
RUINS BY A RIVER
Harrisville

Because most people are incurably romantic about ruins, Harrisville is the glamor exhibit of the Wharton tract. Nowhere have
the Pine Barrens demonstrated more clearly their capacity to ob-
literate man's handiwork than at this ghost town. Egypt's pyra-
mids have survived pitiless exposure to the elements through four
thousand years. Many other monuments to ancient civilizations
remain intact. But in less than a century Harrisville, New Jersey,
has been reduced from a prosperous, stoutly built industrial community to a cluster of fast-vanishing ruins and scattered piles of
rubble.

Romance and mystery abound here. These ruins appear so much
older than they are that some have even compared them with ancient aqueducts. Winding their way under brick arches are curious
canals, now fed only by trickling streams. A constant well rises
from what looks like a stone and concrete coffin. Springs bubble
in deep pits. Amid the crumbling remnants of a gristmill runs the
dry bed of what was a rushing tailrace. The mill itself is thickly
overgrown with vines and brambles. Nearby are gnarled vestiges
of a once thriving orchard. Down a long lane two rows of cellar
holes yawn like so many graves, waiting for the Barrens to conceal
them as they have concealed other cellars not far away. Walkways
which yesterday led to places that mattered today twist through
tangled brush—and lead nowhere. Dominating the entire scene

are towering skeletons of old stone buildings, reaching for the sky
as they crumble toward the earth. All about are the fragrance of
decay, shadows of unremembered things, and the lush, encroach-
ing, overpowering vegetation of the Pine Barrens.

All this, unbelievable as it may seem now, was a community
which was thriving as recently as the 1880's and was called "modern" for its day. No jerry-built place was Harrisville. As a town
it was far different from the collections of frame shacks which had
sprung up near so many of the bog-iron furnaces. The heart of this
community was its paper factory, a vast structure with walls up
to three feet in thickness. The main building was three hundred
feet long and two and a half stories high. It was connected directly
with numerous other structures, most of them also built of stone
and serving key production purposes in what a hundred years ago
was stated to be the largest paper mill in New Jersey. In the cav-
ernous basement of that main mill there were two 13-foot brick
arches which remain intact today. Vagrant streams flow pictur-
esquely through them. The town itself was laid out along a straight
street lined with thoughtfully planted trees, mostly maples. At
night this street was illuminated by gas lamps mounted on tall
ornamental iron posts. Like the mill, some houses were built of
native Jersey sandstone. The owner's mansion was a lofty affair,
and is said to have been panelled in black walnut and to have
housed a gold-plated piano. There was a store, also of stone, a
boxy, two-story structure, the gaunt corners of which still remain

RUINS OF HARRISVILLE (1951)

KEY:
1 Probable Site of Belangee's Sawmill (1760)
2 Probable Site of Isaac Potts's Slitting Mill (1795)
3 Side Canal Cut to Power Gristmill
4 Approximate Line of Onetime Sluice from Gristmill
5 Canal to Boost Power for Paper Mill
6 Cellar Ruin Reputed to be that of Richard Harris Mansion
7 Remaining Cellar Pits of Main Street Homes
8 Main Sluice or Tailrace from the Mill to Wading River

RUINS OF HARRISVILLE

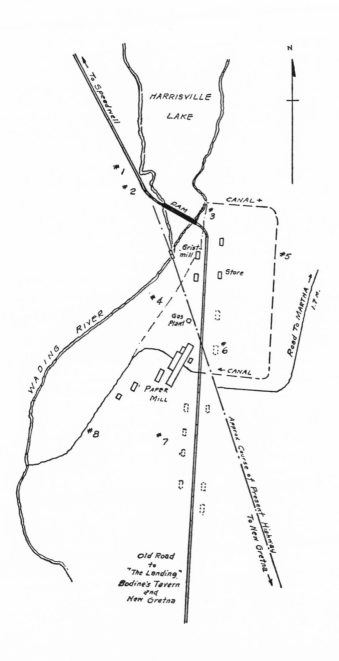

erect. Back in the 1830's this store was earning $3,000 net profit a year; the economics of the village were scarcely small-scale. No doubt farmers from miles around, as well as workmen from various nearby ironworks, made Harrisville's store a "shopping center."

Harrisville lies in the heart of what was an industrial area a hundred and fifty years ago. Eight miles to the north was the busy furnace of Speedwell, once owned by the fabled Benjamin Randolph. Two miles to the east was Martha Furnace, one of the most productive bog-iron establishments in New Jersey. To the west, about eight miles distant, lay Atsion and Hampton Furnaces, and to the southwest were Batsto and Pleasant Mills, with the former's important iron and glass works, the latter's paper mill and other enterprises. Not far to the east was Tuckerton, third officially established port of entry in the United States, with a bustling foreign trade and a busy customhouse. On all sides were sawmills, cutting white cedar and pitch pine; cranberry bogs which had been going strong from colonial days; establishments producing charcoal and turpentine; and the over-dramatized, much-abused "Pineys" gathering sphagnum moss for florists and also roots and plants for vendors of herbs and drugs.

Harrisville is about eight miles northwest of New Gretna, at a point where the modern highway to Chatsworth crosses the east branch of the Wading River. This is often called the "Oswego Branch." In early days the river was the real highway. It is a magnificent sight today as it spills over the broad dam which backs up Harrisville Lake for nearly two miles toward the site of old Martha Furnace. Below the lake the stream deepens rapidly, its amber waters swirling gracefully along majestic corridors which in summer boast an infinity of shades of green. Tidewater is not far distant, and only a few miles farther on lies the sea. Added to these attractions is the fact that the topography of the Wading River basin above Harrisville affords an exceptional source of waterpower; it is not hard to understand why, long before the American Revolution, this lonely locality had a compelling appeal for enterprising men.

The earliest recorded industrial venture in the immediate vicinity was what became known as "The Skit Mill." [1] This included both a sawmill and a gristmill. Built about 1750, it was located on the west branch of the Wading River not quite half a mile above its present junction with the east branch at Harrisville. Early deeds, incidentally, call the east branch *"The* Wading River," and refer to the west branch as "Speedwell Creek." The Skit Mill reportedly remained in operation for 75 years, and some of the pilings used in its dam were visible and were "rediscovered" about twenty years ago. Around 1760 the waterpower of the Wading River attracted a second enterprise, another sawmill. This one was built by Evi Belangee, who dammed the east branch of the Wading River at a point only a few hundred yards above the present Harrisville dam. Estimates are that Belangee's mill was a small one, the dam too, with a head of perhaps five or six feet. No traces of this mill are visible today.

It is the history of most Pine Barrens towns that the lumbercutters were the first on the scene, the sawmill operators next, and the ironmasters third. So it was in Harrisville, which first began to emerge as a community with the construction of "The Wading River Forge and Slitting Mill." A certain amount of mystery has surrounded this phase of the area's development. In his *Early Forges and Furnaces in New Jersey*, the late Charles Boyer mentions the Wading River Forge as "one of the lost forges of New Jersey." It is true that it has been overlooked by historians generally. Study of old deeds and tax records, however, leaves no doubt that the approximate location of this "lost forge" is what later came to be known as Harrisville.

The "Wading River Forge and Slitting Mill" was built by Isaac Potts, "Iron Master of the City of Philadelphia," in the summer of 1795. This date is fixed by an advertisement which Potts placed in the *American Weekly Advertiser* of Philadelphia, for February 11, 1796. In it he offered the Martha Furnace for sale and said that its iron "should command a market near the door as a slitting-mill and rolling-mill has been built last summer." The same advertisement mentions that a lock was "being erected in the slitting-mill dam . . . to admit boats of ten tons burden to pass from the [Martha] Furnace to the tide," thus affording "easy and cheap transportation."

About this time a four-fire forge was built at the same location, that is, the present Harrisville. Both the slitting mill and forge were in full operation when Potts, in 1797, sold the tract on which they were standing. The purchasers were George and William Ashbridge—whose firm was located at the Crooked Billet Wharf in Philadelphia—and Joseph Walker, who also was a part owner of the Speedwell and West Creek Forges. The deed was dated April 24, 1797, and conveyed "the Slitting Mill and other buildings, improvements, etc." [2] Some of the confusion over Wading River Forge undoubtedly arose from the mistaken notion of some writers that a slitting mill was used to cut timber. Actually, it was in common use for fabricating iron, and quite a few such mills operated in colonial times. The old Saugus works had one, and as early as 1750 British iron interests petitioned Parliament to enact legislation which would "prevent the erection in the American Colonies of any Mill or other Engine for the Slitting or Rolling of Iron." At Wading River Forge the slitting mill seems to have been the important unit. After pig iron had been converted at the forge into bar iron, and then rolled into sheets, it was again heated and run through the slitting mill, where rotary shears cut the metal into strips of commercial size.

Nails, in particular, were produced by slitting mills. A nail-cutting machine was invented in 1786 by Ezekiel Reed.[3] This soon made hand-cut nails obsolete. Thereafter nails were cut from nail-rods, which a slitting mill was equipped to turn out. Examples of cut-nails have been found around Harrisville. Their distinctive feature is that the shank tapers on two only of its four flat sides. At first, cut-nails had their heads hammered on by hand, but later, machines both cut the nails and stamped on the heads.

That the Wading River Forge, with its slitting mill, was a good customer of Martha Furnace is clearly established. The diary kept for some years at the furnace mentions various sales of both pig iron and scrap to the "forge and slitting mill." Significantly, when Potts sold the Slitting Mill Tract he was careful to reserve for himself, at nearby Martha, the right to "all Iron Ores whatsoever which may be had or found . . . with full liberty to dig for and hawl away same." The deed also gave him "common use of a lock now completed or to be completed in the Dam" at what is now Harrisville Pond.

In 1805 the brothers Ashbridge sold their interest in Wading River Forge to Joseph Walker, who in turn sold a half interest in the enterprise to John Youle. At some time during this period the old Belangee sawmill appears to have been acquired by Youle and

Walker, perhaps also the Skit Mill, although here the record is unclear. At least, in 1809, they were assessed for four forge fires, one slitting mill, two gristmills, and one sawmill. Youle and Walker jointly operated Wading River Forge for many years, together with Speedwell Furnace and the West Creek Forge. The latter was subsequently called Stafford Forge, which name the locality bears today. All were busy places in those days.

Transfers of interest in Wading River Forge followed, however, and it is recorded that in 1821 Youle's son pledged his interest to Samuel Richards as security for a mortgage of $1,300. This suggests that the Wading River Forge may have been meeting with hard times. It does not appear to have been in operation at all in 1832, when it was sold to William McCarty.[4] One of the title transfers mentions the Harrisville tract as the "Slitting Mill Estate on Wading River."

With the coming of William McCarty, the little community received a new name—"McCartyville." It was given a new direction also. Lean years were at hand for New Jersey bog-iron furnaces generally, and their doom was not far away. Probably McCarty saw what was coming. In any case, he proposed to make paper rather than iron, and formed a company to do so. The capital stock was $65,000, which later was increased to $95,000. By 1835 McCarty had in operation a "double paper mill, two hundred and forty feet in length," with one mill "manufacturing nearly a ton of paper a day." Also, there were a "new stone grist mill, a saw mill, machine shop, blacksmith's and carpenter shops, a stone storehouse, a boarding house, dwellings for the superintendent and workmen," and the store, which he said was turning in a "profit of nearly $3,000 per annum."

That same year—1835—McCarty reorganized his business. He sold his controlling interest to the Wading River Manufacturing and Canal Company, which was incorporated under New Jersey laws. Directors were William McCarty, Thomas Davis, Henry C. Carey, Isaac Lea, and Laurence Johnson. A prospectus shows that the capital stock of this firm was $200,000, of which $25,000 was offered to the public in 1837. This offering was made to finance a "second mill" and also, strangely enough, to underwrite experiments in the production of silk. Cocooneries were becoming something of an industrial fad about that time. As the prospectus also shows, the property then covered 5,000 acres and included a farm of 550 acres, while among the firm's assets was a schooner "that will perform the transportation of goods to and from New York and Philadelphia." [5]

Production of paper by machinery was just getting under way in the 1830's. For six centuries paper had been made largely by hand, and while papermaking machines had been invented around 1800, they were experimental and of limited scope. Linen and cotton fiber, the raw materials of paper manufacture, had always been relatively scarce, and efforts were constantly being made to discover new resources and to utilize substitutes. McCarty, deep in the Jersey pines, found one—salt grass from the marshes of the Jersey seacoast.

McCarty's mill was geared up to produce heavy-grade paper from this salt grass, which was abundant in the nearby areas, including the flats of the Mullica River. The grass was carried from the marshes by barge to "The Landing"—not quite a mile below the paper mill—and then taken the rest of the way by mule team. According to Dr. K. Braddock-Rogers, professor of chemistry, who studied the process, the marsh grass was dumped into cooking vats which were ten feet in diameter and set below the ground level. There were five of the vats, as the plant layout shows.

> Live steam was forced into these vats, which leached out the salts and soluble material. The condensing steam and wash liquors were run off through a tunnel to the sluice. The grass then went to large stone vats fifteen feet in diameter and there was chopped by vertically revolving knives into a pulp. From the macerators the pulp was pumped into a storage tank where it was slowly agitated to keep it from packing . . . and then was ready to be made into paper. In later years a building in which were two drums with knives was built. The drums slowly revolved to cut the grass which was then allowed to drop into another tank. The finish was put on the paper by passing it very slowly over a roller where it was buffed by another roller going in the opposite direction and at considerable speed. [6]

This process produced paper of great strength, which was used principally for wrapping. Some called it "butchers' paper." Also manufactured at the mill were "binders boards and bonnet boards," according to McCarty himself. Some writing paper was produced, and the author knows of a letter from McCarty wherein he states that the paper on which it is written was made in his mill. This branch of the business, however, was not successful, due to a yellowish cast given the paper by the high iron content of the water used in its manufacture. The floor plan of the plant shown here was drawn by the author from a large 1877 survey map of the Harrisville establishment. It locates most of the shops, many of which can be roughly identified among the ruins which still exist.

The use of the word "canal" in McCarty's firm name of Wading River Canal and Manufacturing Company indicates an important aspect of his operations. Waterpower was essential to the enterprise, and as the works expanded, the need for waterpower increased. McCarty had enlarged the old Slitting Mill dam to provide a ten-foot head of water. As his power requirements grew, he launched construction of a remarkable and ingenious system of canals and millraces. There were three of these: first, a canal diverting water from the west branch of the Wading River into the east branch; second, a long, rectangular canal, actually a millrace, to carry water from the lake to the mill, providing water and power for the paper processing; third, a small headrace which operated the gristmill, the tailrace of which broadened into a wide sluice, engineered also to carry away the waste from the paper mill.

The first of these canals was something of an engineering feat. It has been credited to McCarty's successors, but deeds of the period call it "McCarty's canal" and definitely fix the date of its construction during his ownership of the property.[7] The deeds conveyed, among other things, "all right, title and interest . . . in and to the water-power, Tumbling Dam on Speedwell Creek [the west branch of Wading River] above the Mill, and the Canal to convey it into Wading River." This canal, which is clearly shown on many maps, was about one and three-eighths miles long. With its construction, McCarty had gained a six-foot head of water.

McCarty's second canal is supposed to have been dug by hand labor, at a cost of three cents a cubic yard. Its rectangular course can be seen on the sketch, and it covers a total distance of some 750 feet. This canal tapered in width as it neared the mill, from 22 feet at the far end to 12 feet at the mill entrance. Water rushing under the brick arches whirled the great turbines and poured into the massive vats. The sketch shows the outlet, which joins the tailrace from the gristmill, that in turn emptying into the Wading River a quarter of a mile below the mill site. The third and smallest canal supplied the gristmill, which in earlier years seems to have been operated by a smaller millrace. Thus, together with the lake itself, these canals provided McCarty's mill with an efficient system of waterpower.

How much his silk experiments had to do with McCarty's subsequent financial difficulties can only be surmised. It is recounted that he planted a considerable grove of mulberry trees and erected a building for use as a cocoonery. The project failed, and now it is not even known just where the old cocoonery stood. Before he lost control of the property, however, McCarty had completed the main section of the main mill, with its three-foot-thick walls, built the system of canals and a new lake dam, and finished a stone mansion, the foundations of which are still visible.

The name "Harrisville" came from the brothers Harris. There

HARRISVILLE PAPER MILL
Location map of the buildings from a survey of 1877

KEY:

1 Lime Storehouse
2 Storage Shed
3 Machine and Repair Shop
4 Paper Machine Room
5 Boilerhouse
6 Bleaching Room
7 Engines and "Rope Cutter Room"
8 Turbine Room
9 Pulp Finishing
10 Prepared Stock House
11 Rotary Boilerhouse
12 Stuff Chest House
13 Paper Machine Room
14 Gas Plant

HARRISVILLE PAPER MILL

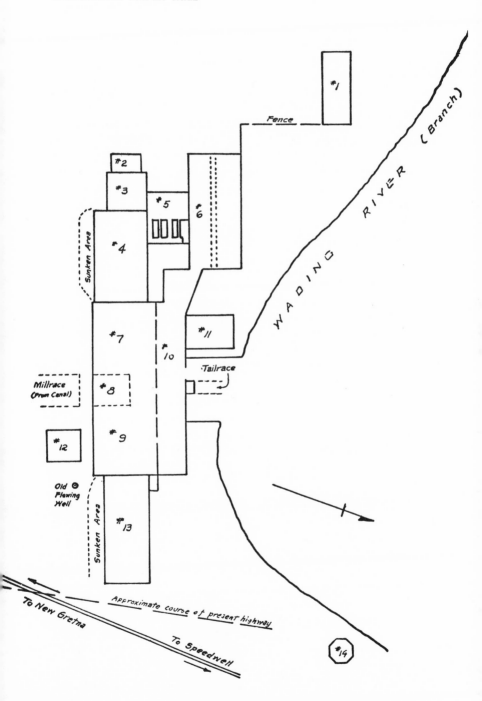

were four of them at first: John W., William D., Richard C., and
Benjamin M. They continued to expand the paper mill and develop the town over a period of approximately forty years. The
records show that in 1851 William D. and Richard C. Harris
entered into a written contract for purchase of the 300-acre property and its paper mill from the President, Directors, and Company of the Burlington County Bank at Medford, Mahlon
Hutchinson, of Philadelphia, and Thomas Albert Haven, of Burlington County. When Haven died before the transaction had
been completed, a court decree became necessary and was issued on
May 24, 1855.[8] That decree carries the following interesting description of this historic property:

> All that Paper Mill, Grist Mill, Saw Mill, Mansion House, Tenant
> house and store houses, with the appurtenances together with so
> much of the land adjacent as is contained in the following bounds:
> Beginning at a stone on the north side of the road leading to
> Tuckerton, thence across said road South eleven and one-quarter
> degrees west, seven chains and eighty links to a stake in the road
> leading to McCarty's Landing, thence South twenty-nine and one-
> half degrees East, seventy-five links to a stone or stake on the
> south side of Wading River, thence up the south and west sides
> of said Wading River the several courses thereof to a stone or
> stake on the West side of said river in the line of the Martha
> Furnace tract, thence across said river North seventy-three and
> one-half degrees East, fifty chains to a small pine sapling in the
> edge of the Pond, thence South eighty-one and one-half degrees
> East, twenty-five chains to a stone or stake, thence South twenty-
> two and one-half degrees West, eighty-two chains to place of be-
> ginning. Containing 300 acres of land or thereabouts recently
> owned by the Wading River Canal and Manufacturing Company.

In 1865 the Harrisville Manufacturing Company was incorporated with a capital stock of $500,000. The directors were Richard
C. Harris, H.P. Harris, C. Heinerman, W. Woodfall, J.S. Fisher,
and A.E. Smith. Richard Harris was the spark plug of the enterprise. He lived in the great house, ran the store, managed the
plant, and even served as postmaster. Harris still further increased
the size of the lake dam, enlarged the canals to gain additional
waterpower, and built both the forward and rear extensions of the
paper mill, together with various other buildings, the remnants of

which are still scattered about. One of Harris' improvements was the introduction of gas. In a small, octagonal building, about 125 feet from the main plant, gas-making equipment was installed in the form of a "Springfield Gas Machine," itself operated by gasoline. This supplied not only the factory, but also the row of ornamental iron street lamps which lined the main thoroughfare. These street lamps made Harrisville the talk of the Pine Barrens towns, and old-timers today recall how their parents and grandparents were awed by its gleaming modernity. The old octagonal gas plant has been something of a mystery to various modern writers, who have declared the structure to be: a springhouse; a blockhouse to guard against Indians; a prison for Confederate soldiers taken captive during the Civil War; a town jail; and an observatory. A survey of the plant, by "A. Hexamer," however, identifies it conclusively as the gas plant.

That survey gives a good picture of Harrisville in its heyday. It is dated August, 1877. Richard C. Harris is listed as owner and superintendent. The main mill then was surmounted by a cupola, while its roof was supported not only by the thick walls, but also by tall iron columns. There was a stone floor in the limehouse, a brick floor in the boilerhouse, and pine flooring elsewhere.

Lighting was provided by the gas plant. Heating was furnished by cast-iron steampipes, and in the paper machine room by a special stove. A small elevator, located in a corner of the main mill, ran from the first to the second floor only. Machinery, all of it propelled by waterpower, included turbines on the first floor, two rope cutters, two rotary boilers, six bleaching tubs, five pulp engines, one "paper machine with calendar," a glazing machine, a washing machine, a lever press, a paper cutter, and a circular saw.

A brick forge was located in the blacksmith shop at the rear of the mill. There were two 200-gallon water tanks in the yard nearby, in addition to a wooden water tank on the roof of the main mill. The latter had a capacity of 2,500 gallons. Fire prevention measures included three fire extinguishers, as well as wooden water casks on each floor of the main mill. Also, inflammable materials, such as oils and wood shavings, were stored well outside the buildings. Similarly, the raw materials for paper making—"old rope, bagging and salt hay"—were stored in piles in the open air "at a safe distance from the mill." As a final precaution, numerous lightning rods were located at various places along the roof of the main mill building.

Harrisville ranked as a prosperous community at that time. There were persistent hopes that a railroad could be built through the town. McCarty had joined the Richardses and others in promoting a line from Camden to Tuckerton by way of Quaker Bridge and Harrisville. That, of course, never materialized. Later the Harrises were associated with efforts to have the Raritan & Delaware Bay Railroad (now the Jersey Central) routed through Harrisville and Egg Harbor on its way to Delaware Bay. That effort failed also, and the line was built some eight miles to the west. As a result, Harris Station was established on that rail route, below Chatsworth, and the paper products of the Harrisville plant were hauled there by mules and oxen over an eight-mile sandy trail through the woods, for shipment to New York.

Due in part to shifts in the streams of commerce which occurred as the railroads expanded, and in part to competing processes of paper manufacture, Harrisville began to decline. An 1877 atlas describes the paper mill as "one of the best and largest in the State," but it also notes that "the village is not as large as it was ten years ago." Richard Harris took great pride in his feudal community, and tales have come down of his rebukes to careless householders and especially to children who let scraps of paper or refuse fall upon the sidewalks. The town was a model of neatness. Harris' brothers appear to have handled the sales end of the enterprise from offices in Philadelphia and New York, while Richard supervised the paper production and ran the community. On the surface, at least, all seemed well.

Below the surface, however, financial difficulties were pressing. Harrisville felt the effects of frequent depressions, and when money became scarce the brothers sought the usual way out— mortgages. Richard obtained $20,000 on one mortgage from Maria C. Robbins.[9] John got another $20,000 from John H. Pratt. With this fresh capital three of the Harris brothers reorganized the business on October 27, 1888. They incorporated the Harris Paper Company with a capital stock of $100,000 (a fifth of the capitalization in 1865!). Howard P. Harris took 375 shares, John W. Harris 188 shares, and Richard C. Harris 187 shares. Then the "new firm" commenced business as of November 1, 1888.[10] On that same date

John Harris deeded his interest in the property to the firm. That deed covered the Paper Mill, with 300 acres, and four other tracts located nearby.[11]

In less than two years the Harris Paper Company was in serious trouble. Among other things it had defaulted on the $20,000 mortgage held by Maria C. Robbins, and she moved to foreclose. The Harrises seem to have made little effort to stave off disaster and save their property. A sheriff's sale was advertised, to take place on February 28, 1891. Advertisements and notices appeared in the *New Jersey Mirror* and the *Mount Holly Herald*.

At the sale, Maria Robbins herself bought Harrisville for a stated price of $10,000. Four days later she conveyed it to Alexander W. Harrington.[12] This deed mentions "all that certain paper mill, grist mill, mansion house, tenement houses and storehouses." Harrington, in turn, gave the lady another $20,000 mortgage. Apparently he was a shoestring operator, and she an optimist.

Not much information is available on Alexander Harrington, with good reason, perhaps, for he did not last long. About a year after his purchase he sold a half interest to Joseph Schneider and Emanuel Ettenheimer, for $9,736, and set up a firm called the New Jersey Manufacturing and Improvement Company. It effected, however, little manufacturing and even less improvement. For the protection of Harrington's creditors, a writ of attachment was issued on March 6, 1893. The writ includes an inventory of the factory at that time: [13] "91 bales of Paper; 125 rolls of paper; 15 tons stock; 150 bales of paper stock; 25 bbls. lime; 17 bbls. alum; 2 casks soda ash; 1000 feet lumber; 7 tons coal; 4 bbls. copperas; 4 tons hay; 200 bbls. of lime at R.R. station."

Complicated legal entanglements followed, but the outcome was inevitable. Harrisville was sold, by William A Townsend, Sheriff, on July 16, 1896. The purchaser was Joseph Wharton.[14]

Well before that, of course, the factory had closed down—forever. Most of the inhabitants of Harrisville had moved away, many of them reluctantly, mourning the loss, not only of a livelihood, but of what they had come to know as "home." True, they were only tenants. All the houses and all the land, as well as the factory and the mills, had been owned by the Harrises and their successors, but it had been a community in which most took pride. Now it was a collection of empty shells.

Worse was to come. In 1914 Harrisville was swept by fire, that great destroyer of the Barrens. Today, with access by a modern highway, the nearest fire apparatus is ten miles away, but even with modern equipment it would be difficult to control a blaze in those parts, once it got started. In 1914 there were only sandy roads. Once the flames gained headway, they raged unchecked until the entire town and much of the surrounding woodland was a charred and blackened ruin. Not a single structure escaped. Building after building fell. It must have been an extraordinary sight when the great mill roof caved in amid sheets of flame, and the mill became, in effect, a monstrous incinerator! Soon nothing was left but a crowd of sandstone skeletons staring out over the lovely lake.

Next came the junkmen, to clean up. Against orders, according to Leeson Small, superintendent for the Wharton Estate, they uprooted and carted off the ornamental street lamps. All remaining vestiges of metal went with them, except the iron pipe of the flowing well, just outside what was left of the main plant. After the junkmen came the vandals. They stole the metal name plate from the recess in the facing wall of the mill. They even drove a wooden plug into the pipe of the flowing well, causing it to burst beneath the ground. When that was repaired, they shot the new pipe full of holes. Finally, when the present dam was built—at a cost of $30,000—Small encased another pipe from the well in a stone and mortar "monument." Other vandals beat down portions of the mill walls and carted away the stone. A comparison of photographs of the main mill ruin taken 25 years ago with those of today show how very much has gone. Apparently only a 24-hour guard could protect the remains of Harrisville from the thieves, butchers, and boors who find perverted pleasure in despoiling not only the Pine Barrens, but the countryside of all America.

Finishing off the obliterating work of the fire, the junkmen, and the vandals are the Pine Barrens themselves. Even stone walls yield to shifting sands and crack under the pressure of infiltrating roots. The green life stirring below the soil is patient in its attack upon man-made structures, but persistent none the less. Unchallenged, the creeping vines tighten their embrace on the ruins. The canals are still visible, but become more overgrown with each succeeding year, just as each twelvemonth finds less and less of the old walls

standing, more and more buried under the verdant avalanche, as
the Pine Barrens slowly reclaim their own. The river alone seems
unchanged and unchanging, save that deer tracks have replaced
human footprints at the old "Landing."

The day may come when Harrisville will be utterly buried, along
with the Skit Mill, the Slitting Mill, the ancient forges, the nearby
furnaces, and the hopes and dreams of so many men. That, of
course, is up to the State of New Jersey, the new owner of Harrisville. It is too early to tell just what is involved in the plans of
New Jersey's Department of Conservation and Economic Development, especially with respect to impounding dams, and possible
flooding of stretches of countryside in order to obtain a great water
reserve for future public needs. Until the day when that is done,
however, the waters of the Wading River will foam along their
curving course to the ocean and evoke for all who see them the
haunting mystery of the Pine Barrens.

7

MARTHA FURNACE

Martha Furnace was built in 1793 by Isaac Potts, a Pennsylvania ironmaster. It was named for his wife, after the custom of those times. This prosperous and productive furnace was located on the Wading River about two miles above the site of Harrisville, where Potts built the forge and slitting mill discussed in the preceding chapter. The Martha tract in Potts's time was bounded on one side by the Tuckerton Road, and the Furnace community provided plenty of customers for the old Tuckerton stages, as did the nearby settlement of furnace workers which, charmingly, was called "Calico."

Isaac Potts & Co. were iron merchants in Philadelphia. Potts himself had operated a sawmill at Valley Forge from 1768 until well after the Revolution, and he also operated the Great Valley Works in Pennsylvania. Perhaps Potts noted the success of New Jersey bog-iron furnaces during the struggle for independence. In any event, he began acquiring land in the Martha area until his holdings reached a peak of nearly 60,000 acres.

There had been a sawmill in the vicinity as early as 1758. This went by the name of "Oswego Saw Mill," and on some old maps its pond was named "Oswego Pond" (not to be confused with the present Lake Oswego in Penn State Forest). Potts purchased the sawmill tract on October 1, 1793, from Samuel Hough, who had operated the mill, and the executors of William Newbold.[1] A year later Potts bought land from John Bodine, presumably the proprietor of the famous Bodine tavern which was located where the

Tuckerton Road then crossed the Wading River. Other tracts were
acquired from the executors of Daniel Ellis.

The date of the building of Martha Furnace is fixed by an entry
in an old account book which Potts kept. It reads: "9 mo. 29. 10
o'clock A.M. Martha Furnace went in Blast. Made the first casting 30th at 3 o'clock in the morning, 1793."

In less than three years a sizeable town had sprung up around
the furnace. Potts himself provided a description of it in an advertisement he placed in the *American Weekly Advertiser*, on
February 11, 1796, when, "desirous of concentrating his affairs," he
offered Martha Furnace for sale, with "about 20,000 acres of land
(pine) and including some cedar swamp."

This advertisement shows that, in addition to the furnace, there
were then at Martha "a forceful grist mill, a saw mill and some
preparation towards erecting a forge." In the community there
were "a number of dwelling houses . . . and a frame house with
cellars, a large log barn, frame coal house, ore houses, the requisite
buildings around the Furnace, a bellows house, bridge house and
moulding rooms . . . large and commodious." Nearby were "near
600 apple-trees and 150 peach trees and several excellent springs
near the furnace."

For supplying the furnace itself there were "wood and ore in
extreme quantity convenient to the works and ores in general rich
and of good quality." One team "is nearly sufficient to bring coal
to the Furnace, and the principle body of ore, lying up the stream,
may be brought by water." Potts also points out that

> The bar-iron should command a market near the door, as a slit-
> ting mill and rolling-mill has been built last summer . . . and a
> lock is being erected in the slitting mill dam by special contract
> to admit boats of ten tons burden to pass from the furnace to the
> tide.

From there, he suggests, the products of Martha Furnace could be
"conveniently directed to any part of the continent or West
Indies."

There is no evidence to show that the lock mentioned was ever
actually built. Rather, the Martha Diary suggests the contrary.
Most of the furnace's products, after 1808 at any rate, were hauled

by wagon to the "landing," which was not far from Bodine's tavern and about three miles from the furnace site.

Over the doorway of the old furnace at Martha there is said to have been an iron plate reading: "ISAAC POTTS, 1793." [2] If so, this plate disappeared long ago. So, too, did Martha Furnace. During the last half of the nineteenth century, what was left of it collapsed into ruins. The remains of the frame village fell down, burnt, and eventually disintegrated. Remnants of the old furnace were sold for scrap iron, and legends tell that hammer heads weighing nearly seven hundred pounds were turned over to junk dealers during the Spanish-American War. Finally the scavengers had their innings, and today there is nothing left.

In its heyday the village had a population of four hundred people. There were forty to fifty houses, a store, a school, a sawmill, a gristmill, and numerous other buildings, even a hospital of sorts. All have been erased from the landscape as completely as if an atomic bomb had fallen there in the midst of the pinelands. Remaining are the lake, now at a much lower level, bits of slag here and there, and a pile of brick and rubbish where the furnace stack stood. To locate even these traces of the past is not easy, since the entire surrounding region is one of utter desolation, and following the main road—a mere sand-track—the visitor is apt to miss the downhill turnoff to the furnace site itself.

It is unlikely that Isaac Potts ever lived at Martha Furnace. His chief business interests lay in Philadelphia. The 1796 advertisement gives his address there, urging prospective purchasers to "call upon or drop a line to the subscriber, Mulberry Street." (Mulberry Street is now Arch.) Potts was quick to lose interest in Martha, since the advertisement for its sale is dated only two and a half years after it commenced operation.

No customers paid hurry calls to Mulberry Street. Nearly five years passed before Potts could close a sale. In November of 1800, however, he deeded Martha Furnace to four men, one of whom had been among the purchasers of his Wading River Forge and Slitting Mill, two miles downstream. The four were John Paul, Charles Shoemaker, Morris Robeson, with whom Paul was associated in business, and George Garrit Ashbridge, of the Crooked Billet Wharf. These men now formed The Martha Furnace Company. They, too, were busy in Philadelphia, and the furnace probably was run, as before, by a resident manager. There is reason to believe, however, that operations at Martha were not very successful at that period, for after four years these new owners were ready to sell in their turn.

In 1805 an advertisement in the Trenton *Federalist* stated that Martha Furnace was to be offered for sale at the Merchants Coffee House in Philadelphia. The advertisement shows that the property now consisted of 15,000 acres, with a blast furnace "as large and substantial as any in the state, in good repair and noted for the soft and good quality of the iron and abundant yield." Also listed are a gristmill, stamping mill, buildings and houses besides well-timbered land with "ore Beds containing Ore sufficient to consume all the Wood." The advertisement further states that the nearby Slitting Mill was a market for pig iron, with another market 12 miles east at West Creek Forge (Westecunk Forge, later Stafford Forge). Moreover, while "Robeson and Paul, 53 No. Water Street, Philadelphia," are named as the sellers, records show that Shoemaker and Ashbridge still held their interest in the ironworks.

When no sale resulted, The Martha Furnace Company got a new manager. In 1806 there first appears the name of Jesse Evans, a man of uncommon character and ability. He was to operate Martha thereafter, through its days of success, and eventually during its decline and fall. Jesse Evans was Washington Township's most distinguished citizen. Aside from his important position as manager at Martha, where he occupied the old Ironmaster's House on a hill overlooking the works, he served officially at various times as Chosen Freeholder, Justice of the Peace, Township Assessor, and Judge of Elections. Evans also was something of an engineer. He made surveys in the Martha area and seems to have directed the building of several bridges. That he was a fair carpenter, as well, is indicated by his making a desk for the schoolhouse.

Jesse Evans and his wife, Lucy, both were Quakers, and an interesting story is told of her devotion to her faith. She had been minister in the Friends Meeting at Tuckerton. When the break between the conservatives and the Hicksites occurred, she joined the Hicksite meeting at Bridgeport (now Wading River), where a meetinghouse was built about 1825. As the Pine Barrens industries declined, more and more members moved away. Lucy Evans remained faithful. She continued to attend regularly, even after she was the only one left, and each first day sat in solitary meditation as minister and congregation.[3]

Martha Furnace finally was sold, on April 6, 1808. On that date Morris Robeson, John Paul, and Charles Shoemaker conveyed their three-quarters interest to Samuel Richards and his cousin, Joseph Ball.[4] Richards owned the Atsion, Hampton, and Weymouth Furnaces. Ball once had owned Batsto, possessed an interest in Weymouth, and was a large investor in real estate as far away as Virginia. George Ashbridge retained his quarter interest in Martha, but after his death it was sold, in 1829, to Samuel Richards.

With outward and visible signs of Martha Furnace practically nonexistent today, it may seem ironic that more is known about this establishment, its village, its people, and their daily lives than about many furnace towns which still survive as shrunken communities. This is due to the existence of the Martha Furnace Diary, a journal kept from 1808 to 1815 by Caleb Earle, who was clerk at the furnace during that period.

Earle, the observant clerk, began his diary on March 31, 1808, with this note: "Finished taking the Inventory of the personal property of Martha Furnace, when Mr. Evans accompanied Charles Shoemaker, George Ashbridge and Samuel Richards to Weymouth, in order to take an Inventory and appraisement of the property there."

The Martha Diary and other sources give a fascinating picture of this colorful community. The focus of activity was the furnace stack; on one side was the millrace, which rushed from the lake to power the water wheel, and on the other the "furnace bank," on a steep hill, this topography making it easy to build a more or less level bridge over to the top of the stack. On the hill was the charcoal shed. Beside it were piles of clam and oyster shells used for flux, and the ore itself, both hauled there by mule and ox-team and soon to be weighed, mixed in proper proportions, carted over the bridge in wheelbarrows, and dumped into the top of the furnace stack.

Below, and to the side of the stack, were the great pair of bellows, also powered by the water wheel, which provided the forced draft in the furnace. (Martha used bellows of the tub type.) Out in front was the molding shed, where the molten metal flowed from the furnace crucible and was diverted into rough gutters, to be made into pigs or ladled into molds for casting various products. Nearby was the thunderous stamping mill, which crushed the fresh ore from the bogs and broke down the slag to recover ore which remained in it due to the inefficient nature of the smelting process. Recovered ore was mixed with fresh ore, then re-smelted.

The products of Martha Furnace included the usual stoves and firebacks, sash weights, sugar kettles, shot, cannon wheels, and various utensils, among them "Youle's Patent Cambosses," which probably were cast-iron cooking pots of some special shape. All these articles were cast in molds, directly from the molten iron, as there was no refining forge at Martha. Much of the pig iron seems to have been sent, at least in the earlier years, to the Wading River Forge downstream, which was equipped to make bar iron, while the Slitting Mill there turned out nails, tires for wagon wheels, and other strip-iron products.

Beyond the Martha Works lay the now-obliterated community. The school apparently was a rough, one-room structure, and the hospital noted in the Diary must have been a crude affair, since there is no mention of a nurse and whenever a physician was required he was summoned from outside—usually "Doc Sawyer," but occasionally a "Dutch Doctor" who never has been further identified.

A little way off was the "residential district" called "Calico." Most of the ironworkers were Irish, although there was a group of Negroes. Living in the village also were the charcoal burners, lumbermen, and ore-raisers who dug the iron out of the bogs. There, too, were some of the teamsters who hauled the ore, charcoal, and shells to the furnace and hauled away the finished products to the "Landing," below Harrisville by Bodine's tavern.

Liquor was a major problem at Martha, as it was at other ironworks of the Pine Barrens. For most of the workmen, the chief sources of entertainment were fishing, hunting, and getting drunk. Since no liquor was available at Martha, its residents were more or less steady patrons of Bodine's, the Washington Tavern of Nicholas Sooy, or "Bucks," which is said to have been on the road from Martha to Chatsworth where it crossed Bucks Run. These extracts from the Martha Diary tell their own story:

1811:   June 22—Richard Phillips discharged for getting
          drunk.

          July 27—William Rose & his Father both drunk &
          lying on the crossway. The old woman at
          home drunk.

          August 20—Peter Cox very drunk and gone to bed. Mr.
          Evans made a solemn resolution any per-
          son or persons bringing liquor to the works
          enough to make drunk shall be liable to
          fine.

1812: December 16—Wm. Rose died very sudden in the coal-
          ing, supposed drunk.

One other diversion was afforded Martha workmen by the
"Training Days." Every male citizen over 18 was supposed to report for military drill three or four times a year. Even during the
War of 1812 this "training" was little more than a joke. Since the
training places usually were nearby taverns, notably Bodine's,
there was often very little training and very much imbibing, the
officers themselves sometimes getting too drunk to "command."
Many of the ironworkers took several days to sober up, and the
furnace meanwhile ran shorthanded. At times, however, it was a
different story. For example, on June 10, 1808, the Diary notes:
"Training at Bodines was very fully attended from ye works. Most
of the hands about the bank retd. orderly." However, save for a
few instances of men leaving to join the Army, the country's military problems affected Martha lightly indeed. After the War of
1812, news of peace did not reach the works until February 24,
1815, two full months after the Treaty of Ghent had brought the
conflict to a close.

Operation of the furnace was, of course, on a 24-hour basis. It
must have been impressive at night to see the bright fire casting its
awesome glow for miles over the surrounding woodlands and hear
the intermittent roar of the blast. Usually Martha Furnace was
put in blast in the spring and blown out the following December
or January. The latter event usually served as a cause for great
celebration. One note in the Diary reads: "Jan. 7, 1809. . . . Blew
the furnace out at 8 o'clock p.m. All hands drunk."

"All hands" amounted to about sixty workers. Most important
of these were the founders, usually two of them, working alternate

12-hour shifts. Under each founder were such skilled or semi-skilled hands as the fillers, or bankmen, who loaded the furnace, the guttermen and molders who handled the molten metal, and the blacksmiths and pattern-makers, besides, of course, the unskilled laborers. Michael Mick was chief founder for some years and did most of the hiring and firing. Many of those fired had a way of getting hired again, and the turnover was less than might be expected, perhaps because the labor supply in such remote parts was none too abundant. There was considerable shifting of personnel among the various jobs at the furnace. The stamping mill, for example, had a wide variety of operators. Some of the skilled men also circulated among the other Richards furnaces and forges. Edward Rutter, a mechanic, "came from Weymouth to put in the Furnace Hearth." Joseph Camp was employed at Atsion as well as Martha. Jacob Wentling (or Ventling), who appears to have been one of the steadiest and ablest men of Martha, also helped out at Weymouth and Atsion.

June 2, 1813, was a dramatic day in the history of Martha. The Diary tells the story briefly: "Great conflagration. The Furnace and Warehouse was this day entirely consumed, but fortunately no lives lost. John Craig got very much burnt."

Luckily the fire did not spread to the nearby town, with its collection of frame buildings, and luckily, too, the forests were not set ablaze. For a while there must have been great anxiety among the workers and their families, but two days later Samuel Richards and Joseph Ball arrived to survey the damage and "concluded to build the Furnace up again." This was done with bricks carted from Speedwell Furnace and "bellowses" and other equipment borrowed from Atsion and Hampton. The task required about two months, and at noon, on August 11, 1813, the rebuilt furnace was put in blast; the report reads, "all going on well."

Since nothing can present Martha as vividly as the Diary itself, extensive extracts are printed in the next chapter. The Martha Diary concludes in May of 1815, when Caleb Earle "started for home"—an odd phrase since Martha had been "home" to him for so long. Isaac Hemingway, Earle's successor as clerk of the works, does not seem to have troubled to chronicle the events of later years. At that period Martha was at the crest of its wave of prosperity. Samuel Richards bought out Joseph Ball's interest in

1822, from the latter's heirs, Ball having passed away the year before. Gordon's *Gazetteer*, published in 1834, describes Martha as follows:

"The furnace makes about 750 tons of iron castings annually and employs about 60 hands, who with their families make a population of near 400 souls, requiring from 40 to 50 dwellings; there are about 30,000 acres appurtenant to these works."

Martha Furnace had lost one of its outlets some years before, when the Wading River Forge and Slitting Mill closed down, and a paper mill was built on the Harrisville site. Now its days were numbered. Like the other bog-iron furnaces of New Jersey, Martha in the 1840's came face to face with competition from the anthracite-fired furnaces of Pennsylvania, with their nearby magnetic ores —competition which it could not meet.[5] It is stated to have been in operation in 1841, when Samuel Richards sold the works to Jesse Evans, its veteran manager. At some time in the mid-1840's, however, the fires were allowed to die out, the water wheel was stilled, the teams stopped hauling, and the plant closed down. In his *Early Forges and Furnaces in New Jersey*, Charles Boyer states that after the furnace was abandoned, "Charcoal was manufactured on an extensive scale and found a ready sale in Philadelphia for domestic and industrial uses. This industry continued until about 1848."

Thus at mid-century Martha had become one more ghost town, and its people, like those of other ghost towns, moved away to seek employment elsewhere. Since the dwellings were all "company houses," there was no problem of homes to sell. There were no leases to break. The inhabitants could just go, and they did, no doubt regretfully.

Jesse Evans apparently sold the tract when the charcoal business petered out. In 1848 it was conveyed to Francis French and William B. Allen, who soon sold it again to Francis B. Chetwood. Martha's last chance for renaissance occurred under the Chetwood ownership. The southward route of the Raritan & Delaware Bay Railroad (later the Jersey Central) was being charted, and since a spur had been laid out from Atsion to Atco, there were suggestions that the main route be turned eastward. Such a route appears on a map of "oswego tract, Formerly Martha Survey, the Property of F.B. Chetwood." The railroad as there "proposed" is shown running through Chetwood, Martha Furnace, and Harrisville, and on to Green Bank, Egg Harbor, and Mays Landing; it is headed for Cape May when it runs off the map. The railroad, when it finally was cut through, passed far to the west, and all hope of reviving Martha as a town perished.

Another interesting feature of this map, which was based on a survey by Samuel S. Downs, of Tuckerton, is the fact that on it the Martha tract is divided into some 34 smaller tracts, all with access to one or more of the trails winding through the property. Some of these "lots" were sold and are still privately held today. The town listed on the map as "Chetwood," incidentally, is shown to be about two miles above Martha Furnace, which put it close to what was reputedly the site of an old brickworks that may have been a Chetwood enterprise. No records of such a venture have come to light. Almost certainly it died an early death.

When Joseph Wharton acquired Harrisville, in 1896, he also bought a large part of the Martha tract. He got little but wilderness. Over the years memories of the old furnace community have grown dim. However, three surviving bars of iron—"pigs"—with the words "Martha Furnace" on them turned up some years ago. Two of them were found, just underground, in the vicinity of the furnace site. Being of bog iron, these "pigs" have not rusted and look today much as they looked in the past. One of these was owned by the late Charles Boyer; another is in the possession of the Wharton Estate; the third is the property of the State of New Jersey. They are practically all that survives of Martha.

8
THE MARTHA FURNACE
DIARY

The Martha Furnace Diary reportedly was found in the office safe at the Harrisville paper factory before that establishment was destroyed by fire in 1910. As far as can be ascertained, two copies were made, after which the original manuscript vanished. Efforts to trace it have been fruitless. Some believe that it was lost in the fire.

One of the copies made directly from the original manuscript was owned by the late Charles Boyer, according to Nathaniel R. Ewan, dean of South Jersey historians. Ewan's own copy was made from Boyer's. A copy of the diary is now in the library of the Camden County Historical Society. At least four other type-scripts exist. One of these is in the library of Rutgers University, another in the possession of Captain Charles Wilson, whose research has been most helpful to the author. The extracts given below were taken from the Ewan and Wilson copies, which were kindly lent for the purpose.

To the best of the author's knowledge, the Martha Furnace Diary has never been printed before, although brief quotations have appeared in Boyer's *Early Forges and Furnaces in New Jersey* and in several magazine and newspaper articles by Ewan, while the diary was source material for the excellent essay, *An Old Jersey Furnace,* by Harvey Moore, published in 1943. More than half the text is printed in the following chapter, and the material deleted is mostly repetitious or of minor interest. The selections have been made with a view to affording a continuous story of the operation of the furnace, as well as the community life, with its pleasures and sorrows, quirks and oddities. The author has attempted to render the spelling of names consistent, since errors in transcription and the fact that some of the entries were probably made by different hands may account for apparent misspellings.

The first entry in the Martha Furnace Diary, kept as we have seen by Caleb Earle, clerk at the furnace, indicates that inventory was taken in anticipation of the sale, by Charles Shoemaker, Morris Robeson, and John Paul, of a three-quarter interest in Martha to Samuel Richards and Joseph Ball, on April 6, 1808. George Ashbridge retained his one-fourth interest until his death, after which it was sold, on May 17, 1829.

The furnace was not in blast at the time of the diary's opening.

GLOSSARY

The following definitions will explain for the reader some of the quaint words and cryptic phrases which constituted the vernacular of the Martha community:

Boarding in the Kitchen: *Taking meals at the mansion.*

Carting Pigs: *Transporting pig iron.*

Clandestine Retreat: *Quitting a job without notice.*

Coaling: *The making of charcoal.*

Hawken (or Hocken): *The town of Manahawkin.*

Hollow Ware: *Pots, pans, kettles, and other open utensils.*

In Blast: *Starting the smelting process.*

Out of Blast: *Allowing the furnace charge to run out and not relighting the fires.*

Landing: *The boat-loading point near Bodine's tavern on Wading River.*

Loads of Shells: *Oyster or clam shells used for flux; not shot.*

Muster: *Childbirth.*

Puff: *An irregularity or interruption of the smelting process.*

Raising Ore: *Digging the ore from nearby streams and bogs.*

Training: *So-called military instruction, usually held at nearby taverns, with more drinking than discipline.*

Stamped Stuff: *Slag crushed by the stamping mill to be re-run through the furnace, to reclaim its high ore content.*

EXTRACTS FROM
THE MARTHA FURNACE DIARY

1808

MARCH

31 *Thursday:* Finished taking Inventory of the personal property of Martha Furnace, when Mr. Evans accompanied Charles Shoemaker, George Ashbridge and Samuel Richards to Weymouth, in order to take an Inventory and appraisement of the property there.

APRIL

1 Saml. Stewart brought 2 lds. Ore in forenoon from Gravelly Point and 1 load from Sassafrass.

6 Perine Applegate come to repair ye Bellows tubs. Teams hauling shells from the Landing.

10 Old cow Peggy found dead in a mire.

20 Perine Applegate has gone home sick.

25 P. Applegate retd. and at work. Arrived Messrs. Ball & Richards.

26 John Lynch helping P. Applegate about the furnace Wheel. S. Stewart hauled 6 loads ore from the Pond.

27 Commenced hauling coal. Stewart hauling sand to the furnace.

28 P. Applegate finished repg. Ye Bellows. Filled up the furnace & put in fire. J. Hedger began to wheel in ore.

29 Peter Cox began to fill the furnace.

30 At 11 O'Clock a.m. put the furnace in blast. Asa Lanning filled 1 turn. J. Hedger banksman. John Cunning doing gutterman's duty.

MAY

1 James Waterford filled 1 turn. Furnace doing well.

3 Furnace doing well.

4 The Furnace made a considerable Puff today. Mr. Evans gone to Tuckerton.

5 Rained most part of day. Luker hauled 1 load ore in forenoon. Stewart 2. Mr. Evans retd.

8 A remarkably gusty day. Furnace made a small Puff.
9 A fire broke out in ye Slitting Mill coaling [at Harrisville] but was got under without sustaining much injury.
11 A very stormy & rainy day. The furnace teams idle. The Furnace made a Puff. No damage done.
12 Phineas Holbert's waggon broke down with the first load ore. J. King repairing it.
13 Saml. Stewart hauled from Leeks in aftn. P. Holbert's waggon mended. Hauled ore in aftn.
15 Everything doing well about the furnace.
18 Finished hauling from Kelly's wharf. Report says James Mc-Gilligan made a violent attempt on the chastity of Miss Durky Trusty, ye African.
31 Moved the ore raisers to Darling's Beds. Owen Hedger & Thos Estell hauled 2 loads ore each.

JUNE

2 Ow. Hedger & Thos Estell brought 1 load ore each from Pappose in forenoon. 1 from Speedwell creek in aftn.
3 A very sultry day. Put the new stampers into N.E. mill. Excellent coal coming in.
13 Teams hauling holloware to the Landing in aftn. Ore hauled in forenoon.
19 Remarkably cold & raw for the time of year. The Furnace made a small Puff.

JULY

1 A very hot day. The Coal Drawers sent in scant Loads. Mc-Gilligan ⅕ short in his second load.
2 Excessively warm. Horses not able to perform more than ⅔ days work.
6 Fine settled weather. Furnace wrought hard last 12 hours.
7 Good iron. Furnace working easier.
10 When Jack Johnson went off this day at noon he bent up one of the hand bars with the intent to spite Richard Phillips, who was absent, and gave the Founder no notice of leaving the Furnace without a Keeper in consequence of which Mick discharged him. N.B. He is charged with damage to the hand bar.
13 Messrs. W. & S. Richards visited the works in Compy. with J. Ball. Dined and retd. for Batsto. A violent storm.
14 The Sloop arrived. Teams hauling from Leeks.

AUGUST

13 Sleeve to the Furnace Wheel gudgeon broke.

14 Stewart's team retd. from Speedwell with new Gudgeon & Sleeve.

15 Jacob Williamson done little work this day. Slept most of the afternoon on the shop bench.

SEPTEMBER

15 Samuel Stewart hauled Shot Moulds to the Landing in forenoon.

19 Jacob Williamson quit work this morning and on the 20th cleared out his A/C exactly balanced without a Cent coming.

21 William Fitzgerald joined work as Blacksmith. The Sloop arrived from Phila. Mary Hedger quit work at the kitchen.

24 Recd. fresh Beef.

25 Furnace made 2 small Puffs this evening.

OCTOBER

6 Stewart hauled Iron in afternoon & hauled the Boat into the Slitting Mill Pond.

8 Messrs. Ball & Richards arrived.

9 Messrs. Ball & Richards went for Batsto this afternoon.

13 Furnace working well. Full lds. of coal. Stock gains on furnace.

NOVEMBER

10 Furnace made a small Puff. Cramer's team carting Pigs.

23 Isaac Cramer's team hauled Moulding sand in forenoon. J. Bodine's team hauled 2 lds. Shells and 1 ton Iron.

27 Mr. Joseph Ball arrived from Weymouth.

28 Mr. Joseph Ball retd. to Weymouth this aftn.

DECEMBER

1 Isaac Cramer's team took a load of Slabs to the woods & 1 ton Iron to the Landing. Furnace working hard.

10 Cramer & Lukers team hauling cannon wheels.

1809

JANUARY

1 Wm. Boner hurt his hand. Jack Johnson filling for him. Mr. Evans gone into the country.

3 J. Bodine's team hauled lds. shells from Walkers heap. Mr. Evans retd. with a quantity of fresh pork.

4 Frost stopped furnace wheel several times.
7 Ore teams hauled hay. Blew the furnace out at 8 o'clock p.m. All hands drunk.
10 Jacob Ventling hunting. Luker hauling hay & Iron.
15 Rained all day. Settled with the Moulders.
16 Moulders started for Phila.
29 Terence Toole made a clandestine retreat from the Chopping. Left his ax, blankets, etc.

FEBRUARY

14 Jack Johnson cut logs in forenoon.
15 Jack Johnson chopping Fire Wood. Sol Reeve in Carpenters Shop. Teams hauled from Hauking [Manahawkin].
22 J. Johnson working abt. the bank [furnace bank].
23 Jack Johnson working at Furnace Hearth.

MARCH

1 Messrs. Ball & Richards arrived. Took an inventory of the stock &c. of Martha.
2 Ball & Richards retd. home.
21 Edward Rutter & Michl. Mick came to put in Furnace Hearth.
30 Carpenters employ'd repairing the Bridge House.

APRIL

10 Patrick Mc Bride commenced as gutterman.
13 A stormy day with sleet & snow. teams all idle except J. King & Co's which hauled 1 ld. ore.
18 Put fire in the Furnace at noon.
20 At 25 m. past 2 o'clock P.M. put the Furnace in blast, Delaney & Cox fillers, Hedger putting in the ore & Donaghau banksman.
21 Teams hauled Moulding sand in forenoon.
23 The Furnace working well.
24 David Jones hauling logs to build a Cabin for the ore raisers at Sassafras & brought back 1 ld. ore. A very wet afternoon.
25 All hands at work.
27 Asa Brown hauled Slabs to Sassafras Cabin. Patr. McBride quit. Rd. Phillips acting as gutterman.

MAY

1 Began to Boat ore. Teams hauling from the Landing.
2 Josh. Hedger and Abr. Cross took the boat to Sassafras. Got her aground coming back and left her all night at the Head of the Pond.

3 Teams hauled ore in forenoon. Corn in afternoon. Stewarts team got off the Bridge at the Furnace Wheel. Not much injury sustained.

4 Sol. Reeve took the Stamping Mill.

12 Mr. Evans went to meet the Stage and brought up sundry dry goods. Men moulding pots and spiders, etc. Jack Johnson took Mrs. Jenkins home in the Sulkey & got drunk.

19 Jno. Bodines team brought 2 loads of patterns from Leeks Landing.

24 Began to mould George Youle's patent Cabooses [cambosses].

JUNE

3 Mr. and Mrs. Evans went to the Beach. Solm. Reeves got drunk.

6 Rained. John Luker carting sand for the Furnace. Iron very high occasioned by stamped stuff.

25 Saw mill took fire. Two new Moulders from Atsion come to go to work.

27 J. Bodine's team began to haul Cambosses.

28 Luker & Stewart began to haul Cambosses. A hard rain with heavy thunder. Coal teams left their wagons out in the coaling.

JULY

12 Metal high in furnace. Hughes lost some castings.

13 Teams brought up sundry goods, flour, soap and sugar from Cramers vessel.

28 Molders all agreed to quit work and went to the Beach.

30 Molders returned from the Beach. J. Ventling drunk and eating eggs at the Slitting Mill. Josh Townsend wanting to fight J. Williamson. Furnace boiled & the metal consolidated in the gutter.

31 Molders all idle.

AUGUST

1 This month begins with good weather. Molders commenced molding for the first time since they came from the Beach.

10 Luker & Stewart carting Cambosses. Molders commenced molding kettles.

SEPTEMBER

19 Michael Woolston come and got J. & Nick Ventling to mold stoves for him. Also took away Widow Bolton's gun.

22 Mr. Evans went to the stage & retd. by Slitting Mill & borrowed some rye.

OCTOBER

4 Teams finished carting cannon wheels. J. Foreman came to purchase corn, Wm. Gifford a scow load of hay.

10 Election at Sooys. Several of the hands went.

24 Stewart hauled a load of corn from Ye Slitting Mill. Sol Reeve frolicking. G. Lungins began to board in kitchen.

29 J. Moore went to Westicunk Forge with Mr. Dickinson. Rained a small shower.

NOVEMBER

1 This month begins with good weather. All hands at their business.

21 Snowed hard all day. Coal teams all idle.

22 Cold night. Froze the pond.

23 A very cold day. Taking up holloware.

24 Snow fell last night about a foot deep. J.M. went to Batsto to get supplies and succeeded. Coal teams all idle.

DECEMBER

1 This month begins with rain. Coal teams went one load each. McEntire helping unload the scow. Recd. all the goods from Capt. Loveland, sa. 3 Hhds. Molasses, 32 Bbls flour, 1 do Sugar, 2 box's Chocolate, 5 bolts Ticklingburg, 15 do Dutch roll, etc.

2 Mr. Evans took McEntire to repair Waden River Bridge.

20 McEntire & Williamson finishing the Kettle Flasks.

31 This evening the Furnace blew out.

1810

JANUARY

1 This year begins with moderate weather. Several of the hands settling & turning in their tools. Jack Johnson started for Philada. Cross driving for King & Co. J. Nash and several drunk. McGilligan ore raising with Taggart. Jas. McEntire securing the patterns.

3 Henry Shinns team hauling ore from the runs. Luker carting logs. Nash stocking the coal since the Furnace stopped.

4 Shinns team hauling ore.

16 Stewart moved the ore raisers. J. King cutting logs. J. Evans went to Batsto.

22 Williamson drunk as a lark.

23 Luker carting logs. Stewart hauling wood & pine to the house.

Owen took a load of boards to the schoolhouse. J.M. fell in the pond.

31 Saml. Richards visited the works.

FEBRUARY

1 Mr. Richards examined the books and retd. to Batsto on his rout to Weymouth. Stewart hauling hay with 3 horses, Brown helping him. Luker carting logs. All the rest teams carting ore from Long Slow. McEntire sick. William Sharp came this evening to survey.

2 W. Sharp surveying. J. King helping him carry chain. Stewart lame and his brother-in-law driving team. Brown helping load hay.

3 Severe snowstorm. All hands and teams idle.

8 J. Evans & King building chimney in the schoolhouse.

14 Mr. Evans making a desk for the School House.

28 William King broke his arm. J. Moore, Jr., went for Dr. Sawyer to set it. Snowed this evening. Lannings children expelled from School for fighting.

MARCH

7 Messrs. Ball & Richards came to take an inventory of the personal property. Ore raisers drowned out & at work on the bank. King sawing. Brown making fence.

9 J. Cramers brought a scow load of shells. Recd. roll of tobacco from Batsto.

13 Town meeting at Sooys [Washington Tavern].

18 E. Rutter came from Weymouth to put in the Furnace Hearth.

19 Luker carting hearth stone ½ day. Rained moderate. J. King helping load hearth stone. Brown helping shingle the Ware House. Rutter dressing the Hearth. McGilligan stocking coal.

20 M. Simons making a frame for the head block of the Wheel.

22 Simons raised a scaffold for a head block for Bellows Shaft.

23 M. Mick finished putting in the Furnace Hearth.

APRIL

1 Sunday. This day rained hard most of the day. Sailed the Boat in the pond.

2 Ore raisers quit on account of the water and went to digging stone.

5 Brown & McMullen laying the bridge house floor. Ore raisers all at work on the bank.

13 Nash filling the Furnace.

14 Nash putting up ore. Mr. Evans put in the ½ charge of Brick in the Furnace. Nash helping put in the ½ charge.

16 Ventling & Brown capping the flume. McEntire sick. Hambleton put fire in the coaling.

17 This morning Michael Mick, Jr., put fire in the Furnace. Ventling & Brown at work on the Cams and Bellows. Sol. Warner & Hunter taking the ware out of the Bridge House room ½ day each.

18 Ventling & Brown working at the Bellows. Terry McCune began to fill. Nash putting up the ore. McGilligan worked ½ hour in the shop and quit.

19 Furnace began to blow 8 o'clock. Cox filling. McGilligan banksman. Shinn & McGilligan doing the guttermans work. Teams hauling molding sand. Ventling at work at the Bellows.

23 Furnace puffed twice today.

24 M. Mick, Jr., house caught fire and burnt down. His wife and boy saved some of the goods. Luker went to move him to his Fathers.

26 Bellows stirrup broke. M. Simons shingling the barn.

MAY

4 Mr. Fugery brot his oval pattern for 9 plate stoves. A hard frost this night.

16 Jno. Bodine ploughed the peach orchard.

19 J. Evans went to Bass River Neck for a mill shaft.

22 Michael Mick's 2 horses team fetched 2 lds. ore from Sassafras.

JUNE

1 This month begins with very cold weather. M. Simons making flasks.

3 This morning about 3 o'clock a fire broke out in the Bellows House & with the greatest difficulty it was got under. The roof & rafters all burnt up & destroyed & with great difficulty the Bellows were saved. Men went to work & the Furnace began to blow before Sunset.

4 H. Shinn drawing nails out of the bellows house roof. J. Evans gone to Bass River on an arbitration.

5 Maurice Simons & Gaskill working at the Grist Mill Shaft. Townsend & Ventling trying to mould the large kettles in sand.

6 Carting all the holloware to the Landing. James Nash employed to watch it. Cramer brot up 2 scow load of ore from North River.

22 Furnace working well. Moulders making plates.

27 Luker & Cox carted stoves to the Landing.

JULY

1 This month begins with moderately cool weather. Furnace making iron fast. Mick making feet for stoves. All hands at their usual employment.

4 Independence Free Day.

6 Sold Forge & Slitting Mill 2 ton scraps.

15 Maurice Simons came today to begin the Stamping Mill.

23 Moses Gaskill began to work at the Stamping Mill.

31 Ventling making casting for the Stamping Mill.

AUGUST

1 This month begins with cloudy dull weather. Furnace making more Iron & better for the Moulders than last week.

11 Moses Gaskill cut off his finger working at the Stamping Mill. A very hot day.

13 P. Cunning working about the Stamping Mill.

15 Maurice Simons finished the Stamping Mill.

16 Patrick Cunning started the Stamping Mill & began to board in the kitchen.

SEPTEMBER

8 Blew the furnace out in consequence of her working bad at about 8 o'clock this evening.

10 Teams hauling iron to the Landing and Hearthstone back.

15 Finished putting in the Hearth about 10 o'clock & put fire in at about 12 o'clock.

17 Furnace went in blast about 8 o'clock in the evening.

OCTOBER

9 Election at Bodine's. Teams to the Landing with Iron.

10 Teams hauling up corn from the Landing. J. Evans at Quaker Bridge.

NOVEMBER

22 Cross & Anderson boating ore.

23 This morning begins with snow. Teams doing nothing.

25 Rain partly all day. Only 2 lds. coal. Shinn & Co.'s team only 1 ld ore from Yellow Cove. E. Rutter moved into Micks house.

27 Shinn & Stewart carting pigs to the Landing.

DECEMBER

6 The men begin to complain of beef. They want Pork.

7 Pond frozen over. Ore boat laying still.

12 Furnace made a puff owing to the wet weather.
21 Rain. Furnace made very bad Iron owing to the wet weather.

1811

JANUARY

1 This month begins with cloudy weather. All hands at work.
3 James McEntire at work in the Saw Mill. Anderson & Mc-
Gahen brought down a load of ore. McGahen fell overboard.
9 James McEntire sawing posts for a stable. We had a very good
sermon preached at the Schoolhouse last evening.
16 Furnace went out of blast. Rain part of day. All in high glee.
26 Teams at work. Saul Warner hauling logs.
27 Mr. Richards sent a drove of hogs containing 103.
29 Finished killing hogs about sunset. Hands kept sober. Cross
driving his team. It was said Cross put his wife out of doors and
told her to seek "lodgins."

FEBRUARY

5 Snowing all day. Teams standing still.
14 Teams all at work. Hands raising the new stable.
28 Stewart & McEntire finished the stable. Teams hauling ore from
Darling beds.

MARCH

2 Milla Townsend made a general muster.
7 Jacob Ventling learning the basket-making trade.
18 Mr. Evans let off the pond.
26 Teams carting ore from Darling beds. James McEntire mending
forebay of grist mill.

APRIL

2 James Hughes helping mend the dam.
5 Hunter, Ventling & Hughes helping at the Furnace.
15 Morris Simons finished dressing the "bellowses."
19 Hunter, Townsend & Hughes taking out the old hearth.
20 Teams all at work. E. Rutter putting things in order for the
hearth in the afternoon.
22 Edward Rutter & Ventling putting in the hearth. R. Allen quit
the mill on Saturday 21st.
25 Stewart began carting coal from Trustys.

28 Jacob Emons went to the Bucks and got very drunk and coming up stopped at meeting to get his sins forgiven him.

30 Geo. Youles sent in his stove pattern. Mr. Evans surveying.

MAY

5 Wm. W. Stockton put fire in the Furnace. Peter Cox filled the Furnace.

6 Saul Warner quit. Cross took his team.

7 Furnace went in blast about 3 o'clock in afternoon. Peter Cox and John Craig filling. Jacob Trusty putting in ore & half Gutterman. Patrick Cunning putting up coals & half Gutterman.

14 Patk. Cunning at the stamping mill. Joel Bodine started with 1500 cwt. of Pigs.

18 Emons hauling sand. Cross hauling moulding sand in afternoon. Garner Grain came after breakfast and quit before night.

19 Great fire in the coaling.

24 All gone well. Ore heap afire. Ore teams carting sand.

27 Cross carting logs for ore cabin. Mary Griffith made a Muster.

JUNE

7 Men at training.

10 Furnace doing very well. R. Phillips stocking coal.

12 Joel Bodine plowed the peach nursery. Teams carting iron and moulding sand.

20 Teams carting ore from Haukin.

22 Richard Phillips discharged for getting drunk. Luker in his place.

25 Milligan & Camp carting iron. They stopped at Bucks. From the juice of the Bucks keg Joseph got intoxicated and let his horse run away.

27 Mr. Ball & wife arrived here about 4 o'clock this afternoon.

29 Mr. Ball started for Phila. Michael Mick had his harvest cut & made all his hands drunk. Joseph Townsend was the only one that returned sober.

JULY

4 Independence. May the name of Washington be immortal and the Federal constitution may it never fail.

6 Hands complained of the rusty pork. Michael Mick reaped the remainder of his rye, women quilting, men a reaping, great hilarity.

10 Mary Luker made a general muster and brot forth a daughter.

12 Grist mill broke. Got the feed ground at the Slitting Mill.

27 William Rose & his Father both drunk & lying on the crossway. The old woman at home drunk. A. Cross making a cabin at Waden Runs.

AUGUST

2 Mary Ducount came out in the stage. Furnace doing well.

8 Moulders all gone to meeting at the schoolhouse.

12 Waters very high. Ore raisers drowned out & working about the bank.

20 Peter Cox very drunk and gone to bed. Mr. Evans made a solemn resolution any person or persons bringing liquor to the works enough to make drunk shall be liable to fine.

24 Ore-raisers digging a ditch through the bank by McEntire's in order to drain the backwater from the mill.

SEPTEMBER

10 Owen Hedger taken with a warrant. Mr. Evans took pity on the poor fellow and kept him from the dungeon.

19 Furnace doing well. All hands at their usual business.

23 The barreling got loose on the shaft which caused them to stop up about six hours. Patrick sent in bad coal.

OCTOBER

3 Furnace making bad iron.

4 Edw. Rutter quit keeping the Furnace on account of her working hard.

8 Election at Bodine's. Men went and gave in their mite and all retd. sober. Edward Rutter off a drinking. It was reported that he got drunk on cheese.

9 Vessel arrived at Leek wharf. Emons brot. a load of corn from aboard her.

11 The Furnace stopped up.

15 Furnace started about 4 o'clock in afternoon. All hands at their business. Rain by showers all day.

16 Teams carting iron. Furnace doing well.

NOVEMBER

18 Mr. & Mrs. Evans went to hear a divine oration at the Bank. Furnace made a great Emission of ore and coal about 4 o'clock in the afternoon.

23 Furnace was stopped up about 12 o'clock this day. Hannah McEntire returned from the country with a cargo of rum. James drunk & sick.

DECEMBER

2 Furnace began to blow this day about 9 o'clock. All going on well.

5 Moulders all returned except George Townsend who I suppose is killing his ox.

11 John Fenimore arrived here with two lds. soap. All going well.

16 Sarah Taylor arrived here with a load of Irishmen all in high glee.

17 Mr. Evans went to lay out lots for woodchoppers at Tranquillity.

19 Furnace went out of blast about 12 o'clock last night. All hands in high glee. Very blustery all day.

21 Jeremiah Fealdon left his coal box catch on fire and burnt his shafter very much, for which he got discharged.

1812

JANUARY

1 James McEntire sawing boards for cabins. All hands at their usual business.

13 Ore-raisers quit and settled.

24 Wm. Hamilton chopping wood. Edward Rutter shoeing horses. Ventling in shop.

FEBRUARY

1 All hands at work. Fine weather in afternoon.

7 Teams hauling from the vessel.

18 Teams carting ore from Sassafras. Jane Hamilton conceived and brot forth a son. Mrs. Core put to bed. Women are all very fruitful, multiply and replenish.

19 Saul Taggart cleaning out the furnace. Stewart carting logs.

MARCH

10 Town meeting at Sooys.

16 Teams took iron down to the landing & hearth stones back.

18 Saul Warner & J. Camp brot each a load of bricks from Hampton brick yards. 1500 bricks.

19 Gene Hunter helping with the hearth. Meeting at School House in afternoon.

25 Michael Mick pulling down part of the bellows house. Edward Rutter finished putting in the hearth.

26 Thomas Andrew & George Hunter cleaning out the Furnace. Isaac Cramer arrived at the landing with a boat & scow load of shells.

27 M. Mick, G. Hunter & T. Anderson helping raise the bellows house.
30 Carpenters began taking down the grist mill.

APRIL

1 Wm. Kenney helping pull down old mill.
3 Edward Rutter started this day for the new Furnace at M. River.
13 Snow all day. Teams lying still. Mr. Richards arrived here this morning.
15 All hands helping to raise the Grist Mill.
20 Men chiefly at training at Bodine's. Rain all day by intervals.
26 Fire put in the Furnace this morning.
28 Furnace went in blast about 4 o'clock this afternoon. James Hearkins putting up the night stock. John Luker wheeling in the ore. Thomas Anderson & John Craig filling.
29 Camp & Stewart carting moulding sand in the afternoon.

MAY

6 Teams carting cinder on the dam.
9 Furnace doing well.
13 Report says Jno. Williams whipped his wife and started for Hanover Furnace.
25 James McEntire brought his daughter home from the Half Moon for fear her morals would be corrupted.

JUNE

3 Great preparations making for Tuckerton quarterly meeting.
12 Ore raisers moved to the new ore bed near the Slitting Mill.

JULY

2 Phebe Craig made a general muster and brot. forth a daughter. Furnace working very stiff.
3 Furnace went out of blast about 6 o'clock this day in order to have the Boshes repaired.
9 Ventling & Mick putting in the hearth. Meeting at the school house.
11 A great battle ensued this day among the Irish.
15 Craig & Anderson filling up the Furnace with coal.
16 Furnace went out some time last night.
20 Furnace made a great puff on Sunday night, but fortunately done no damage.

AUGUST

1 James McEntire gone to Batsto.

5 Furnace making iron pretty fast. All going on well.

10 Rain all day by showers. Ore raisers drowned out. Furnace made light puff.

11 Water very high in the creek. Rain all day by showers.

17 Moulders a deer hunting. George Townsend shot one but did not get it. Mr. Evans surveying. Furnace making bad iron.

21 Mr. & Mrs. Evans gone to Tuckerton.

28 Mr. & Mrs. Tows came out with the stage.

SEPTEMBER

3 Colliers moving to Sooy's woods.

9 Furnace doing well.

14 Coal came in from Sooy's woods.

17 Jane Hamilton was this day tried by the Synod of her church. The crime alleged against her was for using spiritual Liquor, but acquitted.

OCTOBER

1 Teams carting iron to the scow landing.

13 Election at Bodine's. Hands chiefly there. Some very drunk.

15 George Taylor brot down a load of cheese.

30 Moulders out a hunting. J. Townsend killed a deer.

NOVEMBER

1 Last night the paser broke. Furnace stopped up.

2 Teams carting down iron. Mr. Evans looking for a log to make a paser.

3 Furnace started about 4 o'clock in the afternoon.

4 Wm. Hamilton moved to the Slitting Mill. E. Rutter took his house.

6 Camp brot a load of corn from the Slitting Mill.

17 Saul Warner very sick. Attend. by Dr. Fort.

DECEMBER

3 Furnace working bad owing to back water.

5 Emons brot a load of flour & corn from Batsto.

16 Sarah Taylor brot down a load of cheese. Mr. Evans went aboard of the vessel and brot off some goods. Wm. Rose died very sudden in the coaling, supposed drunk.

26 Teams carting iron to the scow landing.

1813

JANUARY

4 Wood choppers very plenty.

7 Furnace went out of blast about 2 o'clock this evening.

8 All hands at their usual business, except the Furnace men.

12 Election at Bodines. Hands chiefly there.

13 McEntire sick after his election frolick. Wm. Penn sawing logs.

FEBRUARY

2 George Hunter taking out the hearth. Ventling in the Smith shop.

9 Mr. Richards arrived here about two o'clock this afternoon. Snow in the forenoon. Teams lying still. S. Taggart gone to Pleasant Mills after wool.

25 Edw. Rutter gone to Weymouth to put in the hearth.

26 Luker helping Maurice Simons hew logs. Stewart pulling down the bridge house.

MARCH

7 Sunday. Snowed hard all day.

12 Teams carting up the hearth stones in the afternoon.

14 All hands helping to put on the bellows house roof. Rain.

20 Teams carting ore from Sassafrass. All going on well.

23 Luker & Trusty putting in the hearth.

26 Camp brot a load of potatoes from Leeks wharf in the afternoon.

APRIL

1 All fools day. J. Luker at meeting in afternoon.

12 William Gibbs & R. Booy filling the furnace. Wm. Mick's widow arrived here in pursuit of J. Mick who she says has knocked her up.

14 Furnace went in blast about 10 o'clock this day.

15 Rufus Booy quit at noon this day. John Craig took his place.

18 Sunday. Furnace made several puffs.

24 Furnace making iron pretty fast.

29 Furnace shaft gave way about 12 o'clock last night.

31 Furnace started about 4 o'clock this morning.

MAY

5 John Page arrived here about noon on a trading voyage. The Yankey came and took his iron plate weighing 4 cwt, 2 qr. 24 lb.

6 John Page started this morning for Batsto.

8 Rudolph Mick fell in the creek this morning and was drowned.

15 Wood wheelers finished a wheeling in the plains this day.

24 All going on well. Mr. Evans started to go to Lumberton.

26 Moulders all idle on account of the patterns not being ready.

JUNE

1 This month begins with very fine weather. All hands at their usual business.

2 Great conflagration. The Furnace and Warehouse was this day entirely consumed, but fortunately no lives lost. John Craig got very much burnt.

4 Ball & Richards arrived here this evening and concluded to build the Furnace up again.

6 Ball & Richards left here this morning for Philadelphia.

7 Moulders clearing away the dirt from the Furnace.

10 Wm. Warner gone to Hampton and Atsion after bellowses stuff.

26 This day the casting house was raised.

JULY

14 Joseph Camp gone to Atsion to keep the Furnace till Martha blows.

15 Capt. Townsend made a draft of the militia at Sooy's.

22 Rimby put in the bellows tubs this day. Maurice Simons framing the bridge house. McEntire staggering drunk. Supposed to have drunk a tablespoonfull more than his allowance.

AUGUST

2 Began to take away the old saw mill.

6 Edward Rutter topping the Furnace stack.

9 James Anderson and John Luker put fire in the Furnace.

10 Ore coming in from Sassafrass.

11 Furnace went in blast about 12 o'clock this day. All going on well.

SEPTEMBER

4 Teams carting ore from below. Furnace doing very well.

15 Mosquitoes very thick.

22 E. Rutter & G. Hunter helping out with Hearth Stones.

24 Teams carting ore from Waden Runs. All hands at work.

OCTOBER

2 Edward Rutter began to keep the Furnace this day at noon.

5 Moulders began to work at George Youle's stoves.

18 Saml. Richards arrived here about noon this day. All going on well.

19 Mr. Richards returns to Philad. All hands at their usual business.

21 Very cold. Began to send away castings this morning.

26 Furnace making iron very fast.

NOVEMBER

3 Furnace doing very well. Meeting at the schoolhouse. Hands chiefly there, it being the day to make their quarterly payments.

5 Mary Trusty made a general muster and brot forth a son. Great joy among the negroes.

9 Iron very sharp last night. Solomon Burr driving Patrick Hamilton's team taking down iron and bringing up shells from the landing.

23 Mr. & Mrs. Evans at Friends Meeting down in Bass River Neck. Furnace making iron very fast.

24 Thomas Taylor and Mr. Evans had a high time respecting some ore that Tommy says was raised on his land.

DECEMBER

1 A fleet consisting of four sail flat bottom vessels arrived loaded with oyster shells.

20 Storm very hard in the afternoon. Teams lying still. Emons carting logs.

22 Furnace wheel stopped about an hour on account of the ice.

25 Christmas. Carters lying still in afternoon. Meeting at Jenkins.

27 Ed Rutter off drunk. Furnace went out of blast early this evening.

1814

JANUARY

1 Bodines team gone to Lumberton.

4 S. Stockton delivered a crop of pork, wt. 4570 lb. All going on in the best order. Luker doing nothing but walk about gentleman like.

5 Snow in the forenoon. Teams lying still.

12 Emons & Reed hauling ore from the Slitting Mill.

13 Mr. Evans started to explore the eastern part of the country and to find a landing for the Iron at the head of the Bay.

FEBRUARY

2 Saml. Taggart finished taking out the hearth.

8 Mr. Saml. Richards & Mr. George Youle arrived about noon.

9 Rain in the afternoon. Mr. Richards and Mr. Youle started for Philadelphia.

14 Six teams started for E. Town point, with each a load of Iron.

15 Saul Warner & George Hunter cutting wood for the Brick Kiln. Teams carting ore from the W. Runs.

25 Davis Cavalier arrived with 8 barrells flour from Lumberton.

MARCH

2 Joshua Earl down surveying at Waden Runs.

3 All hands hauling up Hearth Stones.

21 Jacob Cox & Alexander McKeever commenced filling. Eden Reed began to be gutterman. Put fire in the furnace about 10 o'clock.

23 Began to blow about 12 o'clock this day.

APRIL

1 Captain Townsend was so intoxicated that he did not know the duty of a Captain. At length they had a courtmarshal over him. They brot in for him to pay for a half pint, but refused.

7 Began to mould kettles.

8 Joseph Townsend in the Hospital with the measles.

15 John Youles patterns came today, and a young man to paint them.

MAY

25 Eden Reed filling for Tyler.

27 Patrick McLaughlin began to fill the Furnace.

29 Mr. Evans started to Millville. Hands at their usual business.

30 Mr. Anderson arrived this forenoon with patterns. Mr. Evans arrived home this evening from Millville.

JUNE

1 Edward Rutter started to Millville in likelihood of getting a place.

8 Daniel Reeves keeping [Furnace] in E. Rutter's place. A. McKeever began to fill. Job Cox putting up the ore.

16 Leveling a place for a warehouse.

18 James Donoly & Allen Montgomery boating ore from Sassafrass.

21 Edward Rutter began to work at Blacksmithing and Stamping Mill.

JULY
  5 Mr. Evans and his wife gone to Hampton.
 25 Joseph Hedger hauling stuff for the Boat this forenoon. After-
    noon taking scales & water to landing to weigh pigs.
 30 Solomon Truax & E. Hambleton was married this evening. Had
    a great time, ended of kissing the Bride & others some taking
    gates off the hinges and throwing them in the woods, and some
    to quarrelling.

AUGUST
  2 This evening about ten o'clock Kitty Rutter was put to bed and
    about one o'clock brought forth a fine son, being the first by
    the wife. He was very pleased. Great hilarity.
 23 Has a great fire in the Pines. John Johnston moving Sam.
    Taggart to one of his houses.

SEPTEMBER
 20 Eden Reed started for the Army. Joseph Harvey in his place.
 21 Patrick McLaughlin & James Conley both drunk. John Luker
    filled one-half turn for Conley. E. Rutter also drunk. William
    Warner lame.

OCTOBER
  4 Pat Hambleton & J. Semans gone to Vincentown after stuff for
    the mill.
 24 Johnston hauling logs for Grist Mill.

NOVEMBER
 14 The dam broke this morning. All the ore raisers at work at it.
 15 Mr. Evans & Taggart gone to search for ore.
 18 Mr. Evans & Taggart came back from their journey.
 21 Allen Montgomery began to be gutterman. Job Cox cutting
    logs forenoon. Garner Crane at work in Smith shop. J. Letts
    helping. Joshua Semans making flasks. Johnston hauling logs.
 28 James Bird on his way to New York with the teams. Josephus &
    William Sooy gone to Middletown Point with loads of Iron for
    Mr. George Youle, New York.

DECEMBER
 30 Furnace stopped blowing about 10 o'clock forenoon.

1815

JANUARY
  1 This being a very fine day like spring.

2 Rain part of day. Teams still. Two loads of pork came here this evening.

3 George Hunter & J. Luker helping to cut pork.

17 Rain. Hail. Snow. Cold. I call that a stormy day.

20 We had a meeting in the schoolhouse of friends.

22 Great snow storm which continued all day.

24 Caleb Earl started for the country. Luker after corn & flour. Very bad traveling. Snow very deep.

FEBRUARY

12 William Warner fell out of stable loft, hurt himself so he can't go with his team.

20 Sam Taggart taking out the hearth and boarding in the kitchen.

24 This day we had news of peace confirmed. Teams all standing still. Snowing & raining.

27 E. Rutter came here this morning. His family and goods came on Sunday evening. Pretty fine day.

MARCH

7 Samuel Taggart began with Pat Hambleton's team mending roads.

13 Saml. Taggart & Mr. Evans hunting ore.

16 Geo. Hunter & E. Rutter began to level and fix the hearthstone.

21 Josephus & Nicholas Sooy's team began to work at Martha for T. Evans. The team consisting of 4 mules and one horse, which they are to have 22/6 pr. day.

31 J. Luker hauling cedar poles & clay to furnace in the evening.

APRIL

1 J. Luker to Bisphams Mill.

2 This evening Mr. Henry Smith & Mary A. Taylor was joined in the bonds of matrimony by the Reverend Doct. Fox.

10 John Luker hauling patterns from Carlisle.

12 Charles McGee & Allen Montgomery filling up the Furnace.

16 Began to blow Furnace about 8 o'clock this morning. R. Phillips & James McEntire drunk.

27 Mr. Evans arrived from Philadelphia.

MAY

16 Caleb Earl started for home with his brother. Isaac Hemingway taking his place as Clerk for Mr. Evans.

Ruins of the dam at "Joseph Ball's New Pond" near Batsto. The "New Pond" was new in the 1780's. (*William Augustine*)

Batsto's Company Store. (*Arthur D. Pierce*)

Atsion Mansion. Built by Samuel Richards in 1826. (*N. R. Ewan*)

Martha relics: a "pig" of iron, with the furnace name cast into it; calipers; a kettle; a "frog" doorstop; and an iron cleat, used to reinforce the walls of buildings. (*N. R. Ewan*)

Crosswicks Meetinghouse Stove. Made at Atsion in 1773.

Joseph Ball (1748-1821). Attributed to Gilbert Stuart. (*From The Historical Collection of the Insurance Company of North America Companies, Philadelphia*)

William Richards (1738-1823). By C. B. J. Févret de St.-Mémin. (*Courtesy of Susan Richards*)

Samuel Richards (1769-1842). By Thomas Sully. (*Courtesy of Mrs. Joseph Townsend. Photograph courtesy of the Frick Art Reference Library*)

Benjamin Wood Richards (1797-1851). By Henry Inman. (*Courtesy of Girard Trust Corn Exchange Bank of Philadelphia*)

Jesse Richards (1782-1854). (*Courtesy of Susan Richards*)

Alexander J. McKeone (1851-1928). (*Courtesy of Mary and Raymond Baker*)

Fireback from Etna. About 1773. (*Courtesy of the Metropolitan Museum of Art*)

Fireback made at Batsto to George Washington's order. (*Mt. Vernon Ladies Association of the Union*)

Wood pattern for a Batsto stove casting, discovered in Delaware. (*N. R. Ewan*)

Batsto Mansion, before remodeling by Joseph Wharton. (*Courtesy of Susan Richards*)

Batsto Mansion, after remodeling by Wharton. (*William Augustine*)

Ruins of the Harrisville paper mill. (*Courtesy of the New Jersey Department of Conservation and Economic Development*)

Pleasant Mills paper factory, now a theater. (*William Augustine*)

9
BATSTO

The Indians probably derived "Batsto" from the early Swedes, who used the Scandinavian word "badstu" meaning "bathing place." The white men first applied the name to a gently flowing river—sometimes spelled Batstow or Batstoo—and later to an ironworks community they established nearby. In its heyday, Batsto had not only a furnace, but two glass factories, a brickmaking establishment, two sawmills, and a gristmill. It was a center of agriculture, and vessels were launched there, to sail out the nearby Mullica River to the sea. All that lies in the past. Today, shadowed by surrounding forests, Batsto is a quiet village which wears its historic vestments with dignity and grace.

Charles Read built Batsto Furnace in 1766. The land on which the furnace was erected was purchased in 1758 by John Munrow from the Council of West Jersey Proprietors.[1] Three parcels of land were included in that survey: one "between Atsion River and Batstow Creek"; another, "on the west side of Atsion River"; and the third, "on the east side of Batstow Creek." These three tracts comprise the location of the Batsto plantation, the furnace site, and the village which one finds there today. John Munrow, who lived in Mount Holly, was a land speculator, and the old record books show many transactions in his name. In 1759 he sold a half interest in the Batsto tracts to Vincent Leeds,[2] for whom Vincentown was named. Two years after that Munrow and Leeds sold

Batsto to John Fort, on a time payment basis. Fort ran a sawmill there for several years. He cut much timber, but he couldn't meet the payments, and on May 9, 1764, the Batsto lands were offered at the equivalent of a foreclosure sale by the Common Pleas Court.

Presiding at the "auction" as Associate Justice was Charles Read. The sale was held from noon to 5:00 P.M. Only one bidder appeared. He was Richard Wescoat (or Wescott), who became an associate of Read in the iron business and later was a prominent figure in the Revolution. Wescoat's bid of "300 pounds proclamation money" closed the sale.[3] A deed was given Wescoat jointly by Robert Friend Price, High Sheriff of Gloucester County, and Daniel Ellis, High Sheriff of Burlington County, because the property lay in both those counties.

One year later, on May 3, 1765, Charles Read bought from Richard Wescoat a half interest in several thousand acres of the Batsto property.[4] The price was 200 pounds. Earlier Read had explored the area thoroughly and found that it possessed the four essentials for making iron: bog ore, which assayed high in iron content, and which was abundant in the stream and lake beds; unlimited wood for making charcoal, the furnace fuel of those times; labor, in the nearby squatter community of lumberjacks, hunters, and fishermen at Pleasant Mills; and finally, power, provided by three streams—Batsto Creek, Atsion River, and Nescochague Creek, the power of the first alone being quite sufficient to turn the wheels of the iron-smelting machinery. In addition, there was tidewater just to the east which offered convenient access to river and ocean transport.

Only twenty days after Read's purchase from Wescoat he bought from John Estell rights to "wood fitt for . . . coal and . . . the Ironstone and ore" in a vast tract of land lying along the Atsion River from a point west of Batsto almost all the way to Atsion.[5] This provided the new ironworks with additional extensive sources of both ore and fuel. Read acquired these valuable rights for what appeared to be a pittance. His deed recites that

> The said John Estell in order to encourage the erecting of iron works and for and in consideration of . . . five shillings to him in hand paid . . . doth grant, bargain and sell unto the said Charles Read . . . all the coal wood or wood fitt for the making

coal . . . and also the Ironstone and ore which is now or hereafter may be found on the aforesaid lands, and the tops of the
Saw Logs, always reserving such Timber as is fitt for the saw on
the same, with Ingress and Egress into and through the same
land with horses, Oxen, carriages and servants and the use of
. . . other materials for Coaling and Colliers Houses and for the
Causewaying or Bridging on said Lands.

All this for five shillings? Probably there were private considera-
tions involved, for Estell worked closely with Read in other
ventures, including the establishment of Atsion.

Read's next move was to petition the Legislature for authority
to dam the Batsto River. His political power practically guaranteed
favorable action, and the resulting statute is interesting in that it
gives a clear picture of Read's objectives.[6] Its first section reads:

> An Act to enable the Honourable Charles Read, Esquire, to
> erect a Dam over Batstow Creek; and also to enable John
> Estell to erect a Dam over Atsion River.

> And whereas the Honourable Charles Read, Esq., by his hum-
> ble petition, set forth that he hath proved to demonstration good
> Merchantable Bar-Iron may be drawn from such Ore as may be
> found in plenty in the Bogs and . . . in such parts of this
> Province which are too poor for cultivation, which he conceived
> will be a public emolument; and that in order to erect the neces-
> sary Works, he had lately purchased a considerable Tract of Land
> lying on both sides of Batstow Creek, near Little Egg Harbour
> in the County of Burlington; praying the aid of the Legislature
> to enable him to erect a Dam across the said Creek for the use
> of an Iron-Works; and in order to remove every objection against
> the Prayer of his Petition hath produced a certificate from Joseph
> Burr, Jun., purporting that he, the said Joseph Burr, is and for
> several years past hath been in possession of a Saw-Mill at the
> head of Batstow Creek aforesaid, from whence Boards only have
> been floated down but attended with such Expence as to afford
> a probability that the said Creek will not be hereafter used for
> the like Purpose; hence the said Burr alledges that the Dam over
> the said Creek as petitioned for by the said Charles Read, can-
> not be of any public or private detriment, but on the contrary
> greatly advantageous.

By the end of 1766 Read had built the Batsto Iron Works.[7] The Batsto River had been dammed, and Read estimated that its waterpower was sufficient to operate "four bellows and two hammer wheels." The furnace was located on the north side of the stream, adjacent to a little hillside which permitted a fairly level trestle to be run from the furnace bank to the top of the stack. Nothing of this furnace or its successors remains. The Batsto Forge was erected—not quite twenty years later—a little more than half a mile away, on Nescochague Creek, so as to utilize a different stream from that which powered the furnace.[8]

Little is known of the actual erection of Batsto Iron Works. No records have ever come to light that reveal who supplied the practical skill to build the stack, assemble the bellows machinery, set up the hammers, marshal an adequate labor force, and organize the initial operations. It is known that from the start Read had four partners in his enterprise: Reuben Haines, a Philadelphia brewer; John Cooper, of Burlington, mentioned in deeds as a "gentleman"; Walter Franklin, a New York merchant whose interest was held in trust for his son; and John Wilson, one of Read's friends in Burlington County. Richard Wescoat, of course, was a half owner of the site.

With a swiftness which may well have surprised them, Read's partners soon found the entire iron enterprise unloaded in their laps. Even before the furnace was finished, Read's financial difficulties were mounting. In 1767 and 1768 he sold a half interest in his holdings to Haines, a quarter interest to Cooper, and an eighth each to Franklin and Wilson. This chain of transactions completed, Read had sold out.[9] Soon it was no secret that he was attempting to dispose of all his ironworks interests in the hope of regaining his solvency. Batsto was only the first to go.

The owners—Haines, Cooper, Franklin, and Wilson—probably were bewildered by their responsibilities and appear to have left most matters to a hired manager. A William Doughten is mentioned in that capacity in 1770. Of the four partners, Reuben Haines, the brewer, probably was the man "at the top," if only because he had the most at stake. When Wilson wanted to get out of the Batsto picture it was Haines who purchased his share, on April 27, 1769, and resold it the following day to Joseph Burr. Save for that change, however, the management-by-foursome carried on until the latter part of 1770.

The principal product of Batsto Furnace at that time was pig iron.[10] Some of this was hauled through the woods to Read's forge at Atsion to be refined into bar iron. While no records of the furnace for this period are available, indications are that it was in more or less regular operation. In 1770 it was advertising for a runaway servant, a "lad named Anthony M'Garvey" who wore a "light coloured cloth jacket lined with red, buckskin breeches, old blue milled stockings, strong shoes with strings," and who had "lately been taken out of Gloucester gaol." Batsto's four partners, however, were primarily investors and almost certainly were watching for a chance to sell their ironworks, slough off their responsibility, and if possible turn a profit at the same time. Their opportunity was not long in coming.

Among the more active and ardent yet lesser known patriots of the American Revolution was a Philadelphia merchant and trader, John Cox. During the tense years which preceded the Declaration of Independence he was a zealous worker in the colonial cause. Cox became a member of the first Committee of Correspondence and also of the Council of Safety. Under Cox— later Colonel John Cox—Batsto was to become an important arsenal for Washington and his sorely beset Continental armies.

Apparently the first contact Cox had with the colonial iron industry came about through his marriage, on November 16, 1760, to Esther Bowes, a daughter of Francis Bowes who for a time was part owner of the Bordentown—or Black Creek—Forge. Ten years later, showing considerable foresight, Cox and a partner, Charles Thomson, purchased the Batsto Iron Works. First they bought Haines's interest, on October 12, 1770, for 1,000 pounds. Next day they bought Burr's eighth interest, for 250 pounds, and on October 24 paid 750 pounds for John Cooper's fourth share. They got the remaining eighth share from the Franklin family for 350 pounds, so that the total cost of Batsto in this transaction came to 2,350 pounds.[11]

Burr's deed to Cox and Thomson gives a typical contemporary description of the purchase: "Together with one full and undivided Eighth Part . . . of and in the Buildings, Improvements, Mills, Furnaces, Iron Works, Sluices, Flood-Gates, Free Boards, Mill

Gears, Woods, Underwoods, Timber, Trees, Ways, Waters, Water-
courses, Mines, Minerals, Ores, Metals, Fishings, Fowlings, Hawk-
ings, Hunting Rights, liberties, privileges, etc."

While Cox and Thomson appear of record to have shared
equally in the purchase of the ironworks, Thomson, a leading
patriot in Philadelphia, actually seems to have had but a quarter
interest, which he sold to Cox on September 2, 1773, for 750
pounds. Thus from the start John Cox undoubtedly was "the man
in possession" at Batsto, and he was not long in making his
decisive influence felt.

During the pre-Revolutionary days of the Cox administration at
Batsto, production, previously concentrated on pig iron, was expanded to include a wide variety of commercial and household
articles. This is shown in an advertisement from *The Pennsylvania
Gazette* of June 7, 1775:

> MANUFACTURED AT BATSTO FURNACE in West New-
> Jersey, and to be sold either at the works or by the subscriber, in
> Philadelphia, a great variety of iron pots, kettles, Dutch ovens
> and oval fish kettles, either with or without covers, skillets of
> different sizes, being much lighter, neater and superior in quality
> to any imported from Great Britain; Potash, and other large
> kettles from 30 to 125 gallons, sugar mill gudgeons, neatly
> rounded, and polished at the ends; grating-bars of different
> lengths, grist-mill rounds; weights of all sizes, from 7 to 56 lb.;
> Fullers plates; open and close stoves, of different sizes; rag-wheel
> irons for saw-mills; pestles and mortars, sash weights, and forge
> hammers of the best quality. Also Batsto Pig-Iron as usual, the
> quality of which is too well known to need any recommendation.
>   JOHN COX

With the coming of the Revolution Cox became more energetic
than ever in the colonial cause. According to Boyer's *Early Forges
and Furnaces in New Jersey*, "When the Philadelphia Associators
were organized he was chosen major of the Second Battalion and
later was elected Lieutenant Colonel of the same regiment. In
March 1778, by resolution of Congress, he was appointed Assistant
Quartermaster General." Particularly because he was a member of
the Council of Safety, Cox's ownership of Batsto became an increasingly important factor in munitions production. Once hostili-
ties began, Cox spent much of his time at Batsto. After the British

occupation of Philadelphia he appears to have made his home there. Together with Richard Wescoat and Elijah Clark, he organized the running of the British blockade, the military defenses of the Mullica basin, and the smuggling of supplies overland to Valley Forge.

Even before the Declaration of Independence Colonel Cox had a contract with the Pennsylvania Council of Safety to provide large quantities of cannon balls.[12] Under this contract Cox had agreed to make deliveries by water, but with the effective British blockade of the Delaware, it was impossible to fulfill that part of the agreement. Evidence of the gravity of these problems may be found in a resolution of the Council dated May 20, 1776: "Resolved, That Mr. Owen Biddle be requested to procure 5 or 6 Waggons and send them to Doct'r Coxes' Iron Works in the Jerseys, to bring with all possible expedition the Shot he made for account of this Committee." Two days later Cox himself wrote to Owen Biddle that

> Six waggons are now loaded and ready to start, and I expect will be at Cooper's Ferry [Camden] by tomorrow Evening. My Manager sent off three loads this morning, and I am in hopes that my Overseer, who is gone in Quest of Teams, will return sometime tomorrow with a sufficient number of waggons to take the remainder of the Committee's Order up in the course of next week. You judged well in sending Teams from Philadelphia, it being almost impossible to procure them here at this season of the year, most of the Farmers being busily engaged in planting, and those who make carting a business, all employed in transporting goods from hence to Philada., Brunswick and New York.

There was an interesting "P.S."—"All the Shot ordered by the Committee are Cast."

For this shipment the records show that Colonel Cox was paid 2,481 pounds and 55 shillings. How important Batsto was becoming to the colonial cause may be judged from the fact that on June 5, 1777, Cox—who then also owned the Mount Holly Iron Works—was given a military exemption for his ironworkers. He was, moreover, authorized to set up "a company of fifty men and two lieutenants, with himself as captain," this force to be free of military duty except in case of invasion. The action was taken by

the General Assembly of New Jersey, in a meeting at Haddonfield. That meeting expressed a high opinion of the Cox enterprises, as follows:

> Whereas it is highly expedient that the Army and Navy of the
> United States of America should be furnished as speedily as pos-
> sible with a Quantity of Cannon, Cannon-Shot, Camp Kettles
> and other Implements . . . which the Furnace at Batsto and
> the Forge and Rolling Mill at Mount Holly are well adapted to
> supply. . . .

Cannon as well as shot were manufactured in quantity at Batsto
through most of the war. Even in recent years old cannon balls of
various sizes have been found around the old furnace site, and even
in the lake bed. Batsto in those days also produced large pans for
evaporating salt water, in an effort to get from the sea the salt
which was in much demand for the Army.[13] Colonel Cox himself,
meanwhile, was keeping a vigilant eye on enemy activities along
the coast, and he regularly sent word to the authorities of developments he thought important.

Not long after he was named Assistant Quartermaster General,
Cox appears to have decided to sell Batsto. In poor health, he
probably felt the need to lighten his responsibilities. Besides, he
was now moving into the higher echelons of Revolutionary society.
In his new post he came in close contact with Washington,
Lafayette, Rochambeau, Generals Knox and Greene, and other
notables. Apparently, too, Mrs. Cox and their six daughters had
social ambitions. In any case, after selling Batsto, in 1778, Cox
purchased an old mansion near Trenton, which he called "Blooms-
bury Court," and there he entertained lavishly, and with social
success for the daughters. Boyer notes:

> Rachel, the oldest, married John Stevens, Junior, whose signature
> appeared on many issues of New Jersey proclamation money, and
> who served as State Treasurer from 1776 to 1779. Another
> daughter, Catherine, became the wife of Samuel Witham Stock-
> ton, who was active in the service of the colonies abroad, par-
> ticularly at the courts of Berlin, Vienna and Holland. Mary, the
> third daughter, married Colonel James Chestnut, of Mulberry
> House near Charleston, South Carolina, while Sarah became the

wife of Dr. John Redman Coxe and Elizabeth married Horace
Binney, the famous Philadelphia lawyer.[14]

That Cox made money during the Revolution is indicated, first
by his sharp interest in the privateering trade, for which the
Mullica was a haven, second by the apparently handsome war
profits of Batsto, and third by the fact that when he sold the ironworks to Thomas Mayberry, on October 5, 1778, the price was
40,000 pounds, whereas the cost eight years previous had been all
of 2,350 pounds.[15] Part of this, but scarcely all, reflected wartime
inflation, as did Mayberry's resale of Batsto to Joseph Ball—only
about six months later—for 55,000 pounds.[16]

A few words about Mayberry are in order. He had operated the
Mount Holly Iron Works from about 1772 to 1776, selling the
works to Colonel Cox in the latter year. Mayberry, who seems to
have been "wanted" by the British, then took over Charles Read's
old furnace at Taunton and operated that until 1784, when he
moved to Pottstown, Pennsylvania. Mayberry's purchase of Batsto
had curious aspects. Was it speculation? Or was Mayberry a
"straw man" in an arrangement by which Cox could relinquish
responsibility for operating the Batsto works but continue, as we
shall see, to hold a substantial interest in that property for years to
come? One strange factor in these transactions—Cox to Mayberry
and Mayberry to Ball—is the fact that Ball had long been manager
for Cox, so that under normal circumstances he would have been
likely to purchase Batsto directly from Cox without paying a fat
profit to a middleman. These do not seem, however, to have been
"normal circumstances." They suggest much that remains untold.

Joseph Ball—a Presbyterian and not a Quaker as long supposed
—took over control of Batsto on April 9, 1779. Ball's portrait suggests serenity and strength. He had a clean-shaven, cherubic
countenance with a dimpled chin. His brown eyes seem the more
piercing because of his high, receding forehead; and probably he
was of stockier build than his restful posture indicates. Ball, as has
been noted, was manager at Batsto under Colonel Cox and is
credited with being largely responsible for its heavy munitions
production and the efficient operation of the works under wartime
conditions. At Batsto, Ball laid some of the foundations of administrative experience for his subsequent career as a financier and

corporation executive; and at the old furnace were forged some enduring and distinguished associations, notably that with Charles Pettit which was to play its part in the founding of one of the nation's oldest and foremost insurance companies.[17]

Of Joseph Ball's boyhood, little is known. He was born in Amity Township, Berks County, in 1748. His mother was Mary Richards, sister of William Richards, and his father was John Ball. Joseph's early interest in the iron business probably was fostered by his uncle, as William Richards had been employed at several large Pennsylvania furnaces, and both he and his nephew seem to have become associated with Batsto at about the same time—shortly after Colonel Cox purchased the works. At first Ball operated as Cox's agent in Philadelphia. Later, after William Richards left for military service in the Continental Army, Ball went to Batsto and took over as manager. There he was on close terms with Cox, with his neighbor, Elijah Clark, at nearby Pleasant Mills, and also with another distinguished neighbor, Richard Wescoat. At Batsto, too, Joseph Ball met and won his wife, Sarah Lee, the daughter, by a previous marriage, of Richard Wescoat's wife. Sarah Ball's maiden name has long been uncertain. Some sources have given it as "Richards," which is incorrect. Others have it as "Wescoat" and family records as "May." The name Lee is taken from genealogical data in the Atlantic County Historical Society files, which show also that her mother's maiden name was Brazure, or Brazier. The mother is believed to have been of French extraction.[18]

An interesting confirmation of Ball's residence at Batsto during the Cox regime is an advertisement appearing in the *Pennsylvania Evening Post* for June 26, 1777. It read:

> Wanted at Batsto and Mount Holly iron works a number of labourers, colliers, nailores, and two or three experienced forgemen to whom constant employ and best wages will be given— four shillings per cord will be paid for cutting pine and maple wood. For further information apply at Colonel Cox's counting house on Arch street, Philadelphia, to Mr. Joseph Ball, manager, at Batsto, or to the subscriber at Mount Holly.
> RICHARD PRICE
> N.B. The workmen at these works are by law of this State exempt from military duty.

When Joseph Ball acquired Batsto he was 32 years old, and the Revolution was not yet over. Two outstanding developments marked his five-year ownership of the ironworks: one, the building of the Batsto Slitting Mill, near the furnace; the other, an extraordinary series of financial transactions.

In 1781, the Batsto Forge was built beside what has long been called "The Forge Pond," the southernmost of three roughly parallel lakes to the west and half a mile southwest of Batsto village. This was the site of a once-busy sawmill, built by Samuel Cripps in 1739. It comprised a 12-acre tract which straddled the Nescochague Creek and was prized for its control of the stream at that point. The location of the old forge may still be identified. While the dam has been out for years, the remains of the raceway are visible, and on its banks there are mounds of slag. This is the harder forge slag as contrasted with furnace slag, which is lighter and more porous.

There has been widespread belief that Batsto Forge was built by Charles Read about the time he established the furnace. All recorded evidence is to the contrary. For example, in deeds to his partners Read mentions this tract as the "saw mill." In later deeds to Thompson and Cox from Reuben Haines, and in a 1781 deed from Richard Wescoat, there is mention only of the "saw mill" and "mill dam pond." In 1784, however, an advertisement offering Batsto for sale states: "A new Forge with four fires and two hammers . . . is about a half mile distant from the furnace, on another stream, capable of making 200 tons of bar iron per annum. A Slitting and Rolling Mill with its proper apparatus is also ready for use." This mention of a "new" forge and a slitting mill "ready for use" indicates recent construction, almost certainly under the Ball regime, and probably in 1783. This is also the first evidence that a Slitting and Rolling Mill, capable of producing such things as sheet iron, nails, and wheel tires, was part of the Batsto Iron Works. The capacity of the forge, given as "200 tons of bar iron per annum," equalled the approximate production of Atsion at its peak and was not far below the estimated average output of Ringwood and other large forges of the period. Under Ball, incidentally, Batsto Furnace was still "noted for producing the best hollow ware in America, as well as castings, and pig metal equal to any, and superior to most others."

During this period Ball dammed the Atsion River to form what is now called "New Pond" and appears on old maps as "Joseph Ball's New Pond." It is the second of the three lakes at Batsto, lying between Batsto Pond and the Forge Pond. The dam has been out for some years. No records have yet been found to indicate the purpose to which Joseph Ball put this particular source of waterpower.

Until the close of the Revolution the whole Batsto enterprise, as has been noted, appears to have been highly profitable. Batsto's earnings, indeed, may account for the curious series of financial transactions which were launched shortly after Joseph Ball took title to the Batsto property. Colonel Cox had been made an Assistant Quartermaster General of the Continental armies in 1778, the year in which he relinquished title. Now an arrangement was struck through which another Assistant Quartermaster General, Charles Pettit, and, indirectly, the former Quartermaster General himself, Nathanael Greene, all entered the Batsto picture.[19] By four deeds, all dated October 4, 1779, Ball sold a two-twelfths interest each to John Bayard, Matthew Irwin, Thomas Irwin, and Blair McClenachan. McClenachan was linked with Charles Pettit and also, probably, with General Greene. To his former employer, Colonel Cox, Ball sold a one-twelfth interest. Then on February 12, 1780, he sold a two-twelfths share to Charles Pettit. This transaction may have been the beginning of the long subsequent business association of Ball and Pettit in Philadelphia. More followed. Over the next two years Ball re-purchased the shares of Bayard, the two Irwins, and Pettit. This left Ball with nine-twelfths of Batsto, McClenachan (Pettit and/or Greene) two-twelfths, and Cox one-twelfth. On January 2, 1781, these three bought Richard Wescoat's long-dangling half interest in various Batsto lands, so that after some 16 years the property itself was consolidated. However, the partners cut Wescoat in again the following December, when Ball sold him a one-twelfth share from his holdings.

Well before the close of the Revolution affairs in Philadelphia had been claiming more and more of Ball's interest and attention. He had been friendly with Robert Morris, "financier of the Revolution," and particularly with Pettit. In all probability Ball had met Pettit in the early days of the conflict. As Secretary to New

Jersey's last Provincial Governor, William Franklin, Pettit had
split with Benjamin's son when the latter took sides with the
British. An ardent patriot, Pettit also was a boyhood friend of
General Nathanael Greene, under whom, as already noted, he
served as Assistant Quartermaster General. Pettit is described as a
typical "old-school gentleman of the Revolutionary Era" in "pow-
dered wig, knee breeches, silk stockings and silver-buckled shoes." [20]
No doubt this whole chain of friendships explains at least part of
the fancy financing of Batsto during the Ball regime.

The interest in Batsto of Joseph Ball and some of his friends
waned still further as the post-Revolutionary deflation set in, commerce generally declined, and war industries such as Batsto felt
the pinch acutely. After five years Ball apparently had had enough,
and Batsto was offered for sale, in an advertisement which appeared in *The Pennsylvania Packet* of April 15, 1784. An indica-
tion of the doldrums into which the iron business had fallen is
the fact that the advertisement stresses primarily the attractiveness
of the Batsto estate as a site for gristmills. The furnace was not in
operation at the time, although the advertisement notes that it
"may be put in blast in May or June with a fair prospect of continuing the blast for 12 to 18 months."

This same advertisement gives a good picture of Batsto at that
period. Besides the ironworks, with furnace, forge, and slitting mill,
it was "abundantly supplied with wood and ore of the best quality."
Mentioned are "proper dwelling houses and outhouses . . . in
good order, and an orchard of near two thousand fruit trees." Tidewater came all the way to "the foot of the dam . . . which has
stood for many years, and is well secured against the dangers of
freshets . . . the works [being] seldom injured or delayed by either
ice or back-water." The site, "at the Forks of Little Egg Harbour,"
is located as "38 miles from Philadelphia by land and within one
day's sailing from New York." The river (the Mullica) com-
municates "with the sea by the best inlet on the coast of New
Jersey, is navigable within a few miles of the works for vessels of
200 tons burden or upwards, and those of 100 tons may approach
within one mile; flats and scows may load or unload at the walls
of the mill."

Those interested in purchasing Batsto "are advised to apply to
Joseph Ball, at the said works, or Charles Pettit, in Philadelphia."

How eager the partners were to sell may be judged from the as-surance that payments would "be made easy by taking goods, or giving time for a considerable part of the purchase money." Three months later a "sale" took place, but under conditions through which Ball himself, as well as Pettit and Cox, retained an interest in the Batsto Iron Works for years to come. Further evidence of Ball's continuing interest in this locality is his purchase in 1787 of the nearby Elijah Clark House ("Kate Aylesford" House) and the Pleasant Mills plantation, and his donation in 1808 of the site of the present Pleasant Mills church and burial ground.[21] Ball meantime had moved to Philadelphia. There in 1791 he was elected a director of the Bank of the United States, founded by Alexander Hamilton, and in 1792 he joined Charles Pettit in help-ing to establish the Insurance Company of North America, both men becoming founder-directors and, in turn, presidents of that organization.

Now comes another change and a name which soon became synonymous with the further growth of Batsto and its ironworks, and closely identified with industrial development in southern New Jersey for almost a century. That name is Richards.

At some point during the far-off days between the death of Queen Anne in 1714 and that of William Penn in 1718 Owen Richards arrived at the "green countrie towne" of Philadelphia. With Richards came his wife and four children—three sons, James, William, and John, and a young daughter, Elizabeth. Compara-tively little is known of that family. In 1718 they seem to have settled on a 300-acre tract in the part of Philadelphia once called Amityville, later part of Berks County. For the purpose of this story, the most important information about them is that the son William, who had been born in Wales, sired the William Richards who at Batsto brought prominence and prestige to his family name, and who, in his turn, produced no less than 19 children to carry that name forward.

According to the family record [22] the younger William Richards was born on September 12, 1738. He was about 14 when his father died, leaving the family in poverty. Soon afterward William was sent to Warwick Furnace to learn the trade of an iron founder.

Warwick Furnace had been built on French Creek, in Pennsylvania, in 1737 by the widow of an Englishman, Samuel Nutt. It was a large ironworks when William Richards arrived. It included a bloomary forge, refinery forge, blast furnace, and steel furnace, and was an excellent place to obtain basic training in the iron business. That William Richards did well there is suggested by his marriage to Mary Patrick, daughter of John Patrick, the manager of the works. While Richards was employed there, the Warwick Furnace ranked as one of the biggest iron producers of its day. Its ruins are still visible, and the site now forms part of a lovely estate owned by Joseph N. Pew, Jr.

William Richards was a man of "gigantic mold and great physical strength." He stood six feet, four inches in height and his dynamic quality is suggested in his portrait and in a fine profile engraving by St.–Mémin. By Mary Patrick, William Richards had seven sons and four daughters. After her death, which took place on November 24, 1794, he married Margaret Wood. That was in 1796. She bore him seven more sons and one daughter, "Mamie." The latter, Mary Wood Richards, arrived in 1815, her father's seventy-seventh year.

Just when William Richards first went to Batsto is obscure. It is known that he was active there prior to the Revolution. An advertisement for two runaway Irishmen, dated November 24, 1773, extends the offer of a $12 reward from "John Cox, Jr., in Philadelphia, and William Richards at the works." An advertisement for woodcutters, appearing that same month, advises job-seekers to apply to "William Richards, Manager of said furnace." At that time, however, Richards was still maintaining his home at Valley Forge, and after he left Batsto, about August, 1776, to join the Revolutionary armies, the manager for Colonel Cox was, of course, Richards' nephew, Joseph Ball.

Tradition has it that Richards rose to the rank of Colonel and was in the Continentals' camp at Valley Forge during the cruel and crucial winter of 1777–78. There are no records to confirm the former. It was about three years after the close of the war, and three months after Batsto had been advertised for sale as an ideal spot for a gristmill, that Richards re-entered the picture there. In an "Indenture Quinquepartite," dated July 1, 1784,

Joseph Ball and his partners conveyed the Batsto estate to William Richards.[23] The deed included all the lands owned by Read and his pioneer associates—some 7,000 acres—as well as the rights to the ore and "coaling wood" on the large westward tracts of John Estell. It has been supposed that by this deed Richards enjoyed sole ownership at Batsto. That was not the case. Boyer found in the Swank papers, in the Cambria Free Library at Johnstown, Pennsylvania, an agreement dated July 5, 1784—four days after the deed to Richards—by the terms of which Joseph Ball and Charles Pettit "each agreed to take a one-third share of the purchase." It was specified that Richards was to act as manager of the works and for his services was to be

> provided with out of the common stock of the company all such reasonable accommodations for himself and his family residing at said works, as have been usual on such Occasions and shall moreover be allowed and paid out of the profits or common stock of the company the sum of three hundred pounds annually, as a further compensation for the use of his time and talents in the said Business.

Charles Pettit was named "Agent and Factor" in Philadelphia and was to dispose of the company's output there. One interesting incident during Pettit's agency was his sale of four iron firebacks to George Washington, two of which are now at Mount Vernon, one in the Washington bedroom and one in the West Parlor. It is not known what became of the other two, but all were cast to Washington's special order, with his "GW" monogram, and these as well as some other iron fittings ordered by Washington were almost certainly cast at Batsto.

Legend has it that soon after William Richards took control at Batsto, he had occasion to draw the water out of the mill pond and was surprised to find many tons of pig iron which he succeeded in utilizing before the war prices had entirely collapsed. Yet another legend concerns two rowboats, the remains of which have been seen in the Batsto lake bed, one near the present dam. The story is that when the British were moving up the Mullica basin in 1778 and an attack on Batsto was expected, an order went out to load all the "shott" into boats and sink them. By this means good Continental munitions would be kept from falling into the

Redcoats' hands. But the British were turned back, and the sacrifice—if it occurred—proved unnecessary.

Under Richards the enterprise at Batsto proved increasingly
prosperous. He rebuilt the furnace in 1786 and before long seems
to have paid off his indebtedness, bought out his partners, and
expanded his estate in all directions. Burlington County deed
records show a great number of real estate transactions in which
William Richards was buyer or seller, and no doubt he profited
from realty speculation in those parts, as had so many of his
predecessors.

Richards came into frequent conflict with the Atsion Iron
Works. Lawsuits resulted in at least two cases. One of these involved a canal which the Atsion Company dug between the
Mechesetauxen River and Atsion River above Atsion Lake, to gain
greater waterpower for the Atsion wheels. Lawrence Saltar was
manager of Atsion at the time, with Henry and John Drinker of
Philadelphia as partners in ownership, and this canal was long
known as "Saltar's Ditch." The other major suit, settled by arbitration after many years, involved the old 1765 agreement between
Charles Read and John Estell covering ore and coaling wood
rights in the land along the Atsion River between Batsto and
Atsion.[24] Not long ago a parchment map of this disputed area
was included in a State exhibit dealing with New Jersey iron production. Believed to have been drawn about 1793, it showed that
extensive areas above Sleepy Creek already had been "mined."
Located on the map also are the Batsto forge and furnace, "W.
Richards' Dwelling House," and the Atsion furnace and forge—on
opposite sides of the river from each other—as well as "J. Saltar's
House." Joseph Saltar had succeeded his brother Lawrence as
manager at Atsion.

William Richards modelled Batsto after the iron community
feudalism of Pennsylvania in which he had grown up. He is said
to have "lived like a prince," which is probably an exaggeration.
Boyer is the authority for the statement that Batsto was not
operated by slave labor, as were some North Jersey ironworks of
the time. William Richards, however, had at least two slaves—
Andrew and Ben—whose freedom was ordered in his will. He also
had the usual indentured servants, and an old account book shows
this entry for May, 1811:

Paid for indentured black girl Maria, aged 9, to serve until she arrives at age of 21 years, $105.00. Fee for binding $1.00. Total $106.

In any case, Richards ruled Batsto as a typical lord of the manor. His word was law. Also he was the friend of all workers in need, sold them goods at his company store, usually on credit, and provided medical and legal help when required. So long as he prospered, the iron-molders, colliers, woodsmen, and the rest enjoyed a measure of security. He prospered a long while.

Unlike many men of his day, William Richards thought in terms of retirement. His sons—particularly Samuel and Jesse—had worked with him closely and soon were ironmasters in their own right. Samuel, setting up on his own, acquired Atsion and an interest in Weymouth, Martha, and Speedwell Furnaces. Jesse carried on at Batsto. In 1809 William Richards decided to retire and turned the management of Batsto, as well as the mansion house, over to Jesse. Then he moved to Mount Holly, where he lived the last 14 years of his life.[25] He built no less than three houses there. One of them, an imposing brick structure, lies not far from old St. Andrew's Cemetery where William Richards is buried. He died on August 31, 1823.

A new day had begun to dawn on Batsto even before the passing of William Richards. The United States had been growing and changing. Pioneering was in the air. National expansion was a matter not only of geography, but of new techniques, the search for fresh mineral deposits, and stepped-up commercial competition. All these concerned Batsto. With the approach of the industrial revolution, came increasingly severe cycles of "boom and bust," prosperity and depression, and these were felt sharply even in such a secluded and manorial establishment as this furnace village in the pines.

By the terms of William Richards' will [26] his children were to "share and share alike" in an estate estimated at more than $200,-000. The inventory shows Batsto's furnace and farms were appraised at $55,200, about one fourth the total. Perhaps because either the state of his finances was complex, or because his children already owed the estate over $72,000, his entire holdings were put up at auction in 1824, at the Merchants Coffee House in Philadelphia. Most of the properties were bought in by the family, and Batsto went by deed to Thomas S. Richards, son of Samuel and the Colonel's grandson. Five years later he deeded a half interest to Jesse, who remained the dominating personality, manager of the ironworks, and lord of the manor.

Jesse Richards cut an impressive figure. Always a big man physically, ruddy and robust, with tousled hair, he weighed nearly three hundred pounds in his later years. There was kindliness in his eyes, firmness and determination in his mouth. He is described as "full of enterprise and good nature." Never was there doubt of his ability to carry on for the Richards dynasty at Batsto, and he was in his late twenties when he first took over that responsibility from his father. Jesse Richards' marriage involves an interesting story. The Reverend Thomas Haskins, of St. George's Church, Philadelphia, had married Jesse's sister, Elizabeth. Haskins, previously a widower, had a daughter by his first wife. She was Sarah Ennals Haskins, about six years younger than Jesse. The pastor often preached at nearby Pleasant Mills, and he and his daughter appear to have been fairly frequent house guests at Batsto. So it was not strange that soon Jesse Richards took Sarah Ennals Haskins as his wife. It was a happy marriage, blessed with six children, three boys and three girls. Sarah survived Jesse by some 14 years; she is buried near him in the old Pleasant Mills graveyard.

When Jesse Richards first took over Batsto the bog-iron industry was approaching its zenith in the form of the munitions boom created by the War of 1812. Before long, however, the bog-iron deposits began running low. Either their replacement by nature, theoretically taking place every twenty years, did not keep up to schedule, or else the ore beds were so exploited as to prevent the limonite deposits from accumulating properly in the stream beds. In consequence, the character of iron manufacture soon changed at Batsto, and that in turn brought sharp changes in the related commercial enterprises of the community.

Fortunately the Batsto Store Books have survived and offer a graphic description of life, labor, and changing times in the Richards domain. Not only did Jesse Richards and his family reside at the old Batsto mansion house, but a number of other members of the family lived there at various times. The records show constant comings and goings. Thomas H. Richards, Jesse's son and successor, lived there. Until her marriage Jesse's daughter, Mrs. Bicknell, seems to have been keeping house, and she even stayed on for a time after her wedding. Samuel Richards of Atsion, Weymouth, and Martha, as well as Mr. and Mrs. John Richards of Gloucester Furnace, were regular overnight guests. Clergymen of all faiths, including Bishop Asbury of the Methodist Church, were welcomed and have left accounts of their visits to the mansion. Jesse Richards took many trips, which are duly recorded. Often he travelled to bank at Medford, and he made more or less frequent journeys to Mount Holly, Philadelphia, and New York. Noted, too, is the fact that Samuel P. Richards sailed for Europe on July 19, 1843, a voyage less commonplace at that time than it is today.

Lumbering was always a major activity at Batsto, and there were usually at least two sawmills going. Shiders, lath, and various other rough cuttings were shipped out in great quantity, mostly by boat from the nearby landings along the Mullica River. Various agricultural products helped feed the community, and there were occasional exports of these in later years, including cranberries. However, cranberry culture, under way in the nation's early days (see Appendix IV), seems to have begun late at Batsto. In the 1840's Batsto was making its own brick, and there are notes of brick sales to Weymouth Furnace, among others. In 1851 Batsto bricks were being made by machine. Some charcoal was exported, but, judging from the records, by no means as much as has been supposed. Imports practically all passed through the old Batsto company store. In 1830, for example, the schooner *Confidence* arrived, bearing flannels, prints, ginghams, Irish linen, shirting, brown soap, candles, bedcords, jugs, bowls, flour, shears, knives and forks, table china, and glassware.

Shipping—and also some shipbuilding—were major factors in Batsto commerce. In early years vessels came up as far as "The Forks" of the Mullica. Later on most of the wharfing seems to have taken place either at Mordecai Landing—about a quarter of the way along the road toward Green Bank—or at "A. Nichols" a bit further on, where Nichols kept a store and where the bend in the river, then as now, snuggled close to the road, making an

excellent place for loading and unloading. Some Batsto exports—
especially the lumber and iron pipe—were floated to the ships on
scows, and a channel was dug close to the works to make this
possible. The Batsto Store Books give us the names of some of
these old ships, and often of their skippers. A partial listing follows:

SHIP	MASTER
F. Consul	L.G. Johnson
Batsto	B. Edwards
Gen. Giles	Isaac Post
Samuel Franklin	Edw. Alloway
Argo	O. Loveland
Adjutant	Leek
H. Clay	Johnson
Stranger	
Mary	S. Smallwood
J. Wurts	J. Johnson
Ida	T. Crowley
Rebecca	
Catherine	
J.B. Cramer	
Frelinghuysen	T. Crowley
Martin Van Buren	Risley
Alert	J. Falkingburg
Pearl	John Rose
Tranquil	Isaac Snidegar

The schooner *Frelinghuysen* was built and launched at Batsto
in 1844, the launching taking place on July 18. Presumably it was
owned by Jesse Richards. Only four years before he had financed
the purchase of another schooner, the *Stranger*, for $3,000. Earlier,
in 1837, there had been launched the schooner *Batsto*, also presumably a Richards vessel. Most of these ships plied between
Batsto and New York, Philadelphia, the Rancocas, and Albany,
and there are indications that at least one shipment of kettles was
delivered all the way to Portland, Maine. On their return voyages,
to offset the dwindling supplies of bog ore, they would bring in
pig iron, as well as the usual supplies for the store, the furnace,
and the various mills.

The furnace, however, was the heart of Batsto and the major
source of its prosperity for most of its lifetime as an active indus-

trial community. When Jesse Richards first took control Batsto was noted for the quality of its pig iron as well as the excellence and variety of its castings. Over the years it produced miles of water and gas pipe for major cities. The ornamental iron fence which once enclosed the grounds back of Independence Hall, Philadelphia, was made there, as was the steam cylinder for John Fitch's fourth steamboat. Artistic firebacks were produced at Batsto from the early days, and some were exceedingly attractive. Various historical societies have examples of Batsto products, including cannon balls.

Just when the profitable pig-iron trade died out is not known precisely. It probably was in the latter 1820's, when the bog beds were giving out and other furnaces as well as Batsto were having their troubles in finding raw material to keep going. Many, indeed, gave up entirely when the bog supplies ran low. Batsto survived because Jesse Richards began importing ore, and his mixture of bog ore and "Scotch pig" (the latter apparently obtained via New York) is said to have been in particular favor for some years. Richards, however, seems to have taken his iron ore pretty much where he could get it. On November 8, 1831, note is made of a consignment of "60 tons Morris County ore @ $5." The following December 31 the *Confidence* brought in 50 tons of "Staten Island ore @ $3." Other shipments listed were "Schuylkill ore" and "Kincory" ore, while overland "128 tons of ore from Rancocas" were hauled in June, 1840, and 271 tons in October, 1841.

Jesse Richards rebuilt Batsto Furnace for the third time in 1829. The Store Books give many details concerning that third Batsto Furnace. In 1834 it loaned Martha Furnace "1 set birch head patterns" and "one birch gudgeon." In July, 1835, it loaned Gloucester Furnace, another Richards enterprise, 94 gallons of molasses. In 1840 it sold Martha "1 pattern @ $3.50."

Particularly interesting is a notation in the books for February 24, 1835, giving the dimensions of a new furnace hearth which had just been installed:

> Dimensions of the Hearth put in this day at Batsto Furnace are 19 inches at the Bottom of square—22 inches at top, beveled to 24 inches. Hearth 4 feet 11 inches high . . . cut 12½ inches from bottom, 9½ inches from back wall. Boshes put in 9 inches from Bevel of two inches in hearth to the foot.

Two years later another new hearth was described:

Bottom Hearth to top of square	5 ft.		
Across bottom Hearth	1 ft.	7	in.
Across head Square	1 ft.	9	in.
Height of tuyere	1 ft.	1	in.
Bevel of boshes		9½	in.

In April, 1835, after completion of the first-mentioned hearth, these notations appear:

April 5—Burden put on furnace at 12 P.M.
April 6—Gate drawn this day at 12 P.M. The first iron was put
    into castings.

Until December 12 of that year—1835—the furnace ran full time. It made "250½ days" with 30 to 36 men attending it. For December 12 there is the note: "Blowed out at 10 p.m."

The Store Books indicate that Batsto was not as hard hit by the depression of 1837 as were many other enterprises. The most important development of the later 1830's, however, took place on August 20, 1838; it is noted that the furnace "stopped at 12 o'clock to put up hot blast." This suggests that the Richards brothers kept abreast of the times, for the hot blast method of smelting iron was only then coming into general use. Atsion had installed the new hot blast sometime earlier.

Despite these measures, the hot blast method of smelting meant the beginning of the end for Batsto and other charcoal-burning furnaces. By facilitating the use of coke instead of charcoal, it gave Pennsylvania's iron and steel industry, with its closeness to the coal mines, a competitive advantage which never was to be overcome.[27] This, of course, was far from obvious at the time. Batsto then was still doing a tremendous business, especially in gas and water pipe. One consignment alone, on November 16, 1840, to the Manhattan Gas Light Company, included 4,410 feet of 6-inch pipe, 2,619 feet of 4-inch pipe, and 2,700 feet of 3-inch pipe, for a total, with "19 branches," of $5,424.53. Further invoices for pipe mention as purchasers, among others, the Atsion Iron Works and Burlington County.

As each year passed of the critical 1840's Pennsylvania competition made itself felt more keenly in Batsto. Unlike many iron-masters, who hastily abandoned their enterprises, Jesse Richards was determined to fight back. Perhaps he realized that the old-style furnaces were on the way out. In any event, about 1841 he built an entirely new one—a "cupola," or re-smelting furnace, which refined pig iron into finished products of higher quality than had previously been possible. A second cupola was built in 1848, but its debut was inauspicious as is shown by these Store Book entries for the year:

April   17—Cupola started this day.
April   19—Cupola not in blast on account of storm.
April   22—Cupola not in blast.
May   17—Cupola idle for want of charcoal.
June   17—Cupola idle. Too hot and stopped up with cinders.
June 21–26—Cupola idle for want of pigs.
July   12—Idle. Putting inside walls in cupola.
August   12—Cupola idle. Pipes burnt out yesterday.

Troubles with the cupola persisted. In November it was repaired again, and new walls installed. In December of 1848 there appears this notation: "Cupola idle for want of iron. Men working in blacksmith shop, salting pork and odd jobs. Iron business bad." During the next four years operation of the cupola continued to be irregular due to "want of" something—coal, pig iron, or customers. In February, 1852, for example, it is noted by Richards' New York agent that "business is exceedingly dull."

That fact may have had some connection with a recently discovered record of the cost of iron-making at that time. Dated Batsto, February 26, 1852, this "Statement of 1 Melt of Iron" offers these interesting figures:

4½ tons Pigs: 2¼ tons Am. & 2¼ do. Scotch	
@ 21 & 23 pr. ton	$99.00
Freight, including scowing	5.40
Commission on purchases 2½ pc.	2.61
18 Cwt coal	4.50
Moulders wages	22.44
Cupola man and filler	2.67
Carpenters wages and ½ time of blacksmith	2.03

Machinists' pay and materials	3.00
Horse and cart pr day 1.50; proving pipes 1.20	2.70
½ cord wood to dry cores 1.25; ¼ cord shiders	
to start cupola .39	1.64
Carting sand and clay for day	1.50
Cost of hay, nails and tools . . . say . . .	1.00
Freight to N.York on 4 tons @ 1.50;	
down river @ .20 per	6.80
Commissions on sales . . . say . . .	7.50

Cost of one melt   162.79

The above will produce:

12 pipes 6 in. dia. 300 lbs. ea.	3600 lbs.
29 " 4 " " 160 " "	4640 lbs.
5 " 3 " " 120 " "	600 lbs.
	8840 lbs.

The Cost of making
1 Pipe 6 in. dia. is $5.52½ or 61½ cts pr. foot
1 " 4 "   "   "   2.94¾ or 32¾ "   "   "
1 " 3 "   "   "   2.21   or 24½ "   "   "

In May of that year, 1852, an entirely new cupola was constructed, but its operations were spasmodic, as before. It solved no problems. While the earlier cupolas were running, the old furnace itself had continued in operation. In 1847 its laggard pace was recorded by a note that it went into blast at 11:00 a.m. on September 21, was "stopped up by back water" five days later, resumed work on September 29, and so on. On January 25, 1848, this fateful notation was made: "Furnace blew out at 9 o'clock p.m." That was the last record of old Batsto Furnace.

There is a Chinese proverb: "Dig a well before you are thirsty." Jesse Richards possessed such foresight. He undoubtedly saw in advance the day when men would cease to make iron in the pines. He was determined that Batsto should survive, and to that end he set up another, entirely different, industry—glass manufacture.

The building of Batsto's first glass factory—in later years there were two—was completed, and the first glass blown, on September 6, 1846. This was before the cupola was constructed and while the old furnace was still in action, albeit intermittently. Much of Batsto's glass output went the same way as its iron: to New York, Philadelphia, and other large cities—especially New York. The Batsto Glass Works specialized in making window glass and other "flat glass," and enjoyed an extensive trade in the glass panes then in great demand for municipal gas lamps. There were four grades of Batsto glass. In 1848 they were named:

> 1st Quality: Union Extra.
> 2nd Quality: Union First.
> 3rd Quality: Greenbush Patent.
> 4th Quality: Neponset Patent.

More elaborate names were bestowed a little later on, as follows:

> 1st Quality: Star & Moon.
> 2nd Quality: Jesse Richards.
> 3rd Quality: Sterling.
> 4th Quality: Washington.

As far as can be determined, the first Batsto glass was shipped to New York on October 3, 1846, on the schooner *Henry Clay*. Thereafter glass shipments were frequent. Where possible the vessels brought back pig iron, but as the years passed more and more of them came back "in ballast." The glass business presented serious problems. Fires seem to have been fairly frequent and put the glass house out of operation for varying periods, some of them extensive. In 1851 the following complaints were made concerning "deficiencies in Batsto glass":

> 1: Cut too much short of measure.
> 2: Corners broken off lights.
> 3: Not uniform in selection of the different qualities.
> 4: Badly packed—sometimes not a particle of packing in the box.
> 5: Large glass blown too thin and not well flattened.

Batsto's second glass house was built shortly after the fires of the old iron furnace died out. An entry for February 23, 1848, reads: "Raised new glass house." The following June the old glass house burned, and it was not back in operation until November. On

January 29, 1850, the new glass house went the way of the old. It is noted that the "new glass house burnt down this morning at 2 o'clock." Yet another fire took place in May, 1851, but by September of that year both glass houses appear to have been back in operation. So it went over the years. The wonder was that these fires did not sweep the entire town, fire-fighting measures being as inadequate as they were in those times.

Batsto's greatest disaster struck in the summer of 1854. The Store Books tell that story simply:

> June 17—Mr. Richards demise 8 o'clock a.m.
>   18—All the work on the premises stopped from 16th to 21st on account of the death of the proprietor.

Jesse Richards' tombstone, hauled from Mount Holly the following October 25, bears the inscription: "Beloved, Honored, Mourned." All that he was, indeed, not only in Batsto, which so long had revolved around him, but in the wide circle of his friends, relatives, and associates throughout the eastern seaboard states. Not only was Richards outstanding for his skill as an ironmaster, his drive as a business executive, and his grasp of commercial detail, but he had established many and valuable contacts. He had agencies in both New York and Philadelphia. He had served in the New Jersey Assembly from 1837 to 1839. That he kept in close association with other ironmasters is suggested by a $50 subscription in 1831 to the "Iron Masters Fund"; $50 was not small change then. Above all, Jesse Richards knew how to skate over the thin ice of nineteenth-century finance, which usually was a lot thinner than the winter ice on Batsto Pond.

It is commonly supposed that Jesse Richards owned Batsto outright, and that he was extremely wealthy. Neither impression seems quite accurate. As previously noted, Thomas S. Richards long possessed a 50 per cent share in Batsto, and on May 12, 1845, after his death, notices were put up at Pleasant Mills, Mays Landing, Port Republic, and Weymouth announcing the "sale of ½ of the Batsto property belonging to Estate of T.S. Richards, Dec'd." In order to purchase this half interest of his late nephew, Jesse appears to have borrowed the money through a bond issue which was secured by a first mortgage on Batsto. A note to this effect appears in a letter from one of Jesse's brothers to Robert Stewart, manager at Batsto. The mortgage, for $13,425, was given in 1845, and its foreclosure some thirty years later was to mark the end of the Richards era at the old plantation. About the same time as the bond issue, Jesse Richards took into partnership a James M. Brookfield, and the firm name "Richards & Brookfield" appeared in the company's records. This association lasted less than three years. In a financial statement on its dissolution, in 1848, there is no mention of iron or ironworks interests, but listed among the assets of Richards & Brookfield were "Goods in store, Materials, Fixtures, Tools in factory, Horses, Mules, Waggons . . . wood on hand, glass on hand," besides balances due of $1,200.58 and cash on deposit in the Burlington County Bank at Medford amounting to $158.75.

This latter figure of $158.75 is significant. It reveals Jesse Richards' close figuring in financial matters and suggests that most of the Batsto community business was transacted with a minimum of cash changing hands. The Richards cashbook from 1845 on mentions numerous trips to the Medford Bank, with sums being borrowed, paid, and borrowed again at very frequent intervals.

In 1846–48, Richards owed money to several of his employees. Listed are notes for $544 to J. Peterson and one of $200 to F. Fralinger. Probably these sums were for wages, a situation which shows the economic inbreeding of the community. In turn, however, there was credit for all at the company store. Houses were leased at nominal rentals. Board is noted to have been "two dollars a week." With few occasions for spending money on entertainment, there was no pressing need for cash. It was possible to manage remarkably well without it.

A glimpse of the social life of the community may be of interest. Aside from church activities at nearby Pleasant Mills, there were only occasional "events." In January of 1845 the Furnace "paid for sundry persons to see a Magic Lantern." Cost was $2.12½. On November 7, 1850, there was a "circus show at J. Wilson's." The following April 21 a "show of beasts at Pleasant Mills" resulted in "one-half day of hands lost." Otherwise, aside from town meetings, hunting, fishing, and bouts with liquor, life was tranquil, and spent chiefly in long days at the factories, and short evenings at home.[28]

Stagecoach service appears to have been fairly regular. The

stages were often used by the Richards family, and they also brought occasional visitors both to the Big House on the hill and the little houses in the village. In those days stagecoaches did not confine their activities to transporting people. Once the driver unloaded two dozen shovels at Batsto. On other occasions he was "freight agent" for "old saws," 138 yards of flannel, cotton hose, and sundries. Consigned to the Big House in September, 1841, was "One box Burgundy Claret: 24 bottles, per stage $6.50." Further shipments by stage consisted of medicine, a half-chest of tea, two dozen "Buckskin mits," and "2 doz. spelling books."

Another factor in the activity of those days was the taverns. These not only served the stagecoach routes as stopovers and refreshment stations, they also functioned as community gathering–places. Quaker Bridge, today as desolate a spot as the Pine Barrens can offer, was then a favorite place for elections, town meetings, oyster suppers and other festivities, for which the tavern there seems to have provided excellent accommodations. Jonathan Cramer's tavern at "The Mount" was a particular haunt of Batsto folk, and was also a polling place in later years. Most famous of all, as has been told in an earlier chapter, was the Washington Tavern, about five miles from Batsto, which was conducted at that period by Paul Sears Sooy. Still others were the Pleasant Mills Tavern of Samuel Kemble, Bodine's near Harrisville, and the Nichols place near Crowleytown. All these are gone today. Of most, scarcely a trace remains in the ghostland in the pines.

Jesse Richards' death meant to Batsto the loss of its vital force. All the mills and dams and machinery were still there, all the woods and waters, the workers and their kin, the Big House and the little houses. The dynamic energy which Richards possessed died with him. He could not pass it on. By a will drawn just before his death, Jesse Richards bequeathed his estate to his wife, his three sons, Thomas Haskins, Jesse, Jr., and Samuel P. Richards, and three daughters, Elizabeth, Ann Maria, and Sarah. The estate was to be divided "according to the method provided by the statute of the State of New Jersey." These heirs, with Thomas H. Richards in charge at the Big House, strove to carry on. Despite their efforts, however, the whole tempo of Batsto slowed down until, finally, all motion ceased and community life withered away.

Thomas H. Richards was a cultured, amiable man who was

well-liked and well-intentioned. He had not, however, been trained in that hard school of experience from which both his father and grandfather had graduated, nor was he the financial figure skater his father had been. Worst of all, he was cheated out of thousands of dollars by one of his agents. Economic conditions became more unfavorable, assets began to shrink, and the end of Batsto as an industrial center was written in the stars as clearly as it was written in the accounting ledgers.

During 1855 the cupola worked only 26 days—16 days in March, 3 in April, 7 in May. In July the one operating glass house shut down, but reopened in October and finished out the year. In 1856 and again in 1857 the cupola did not operate at all. On August 3, 1858, comes this fateful notation: "Tearing down Cupola."

The Batsto Iron Works had lasted all of ninety years. It had served the new nation well. It had brought poverty to its founder, wealth to many others. It had fought hard to survive against adverse conditions, but in vain. Now that chapter at Batsto was closed. Two more remained to be written.

Even with the furnace closed down, hope lingered in Batsto. If no one wanted its iron, customers still were to be found for glass and lumber. Thomas H. Richards sought to concentrate on those commodities. Troubles, however, continued to pile up. On June 13, 1856, a fire closed down the glass factory. Three weeks later one sawmill broke down. From July 11, 1856, to September 8, 1857— well over a year—the glass house was idle. In 1858, however, the glass factory had been rebuilt and enjoyed a full year's operation, a feat which was almost equalled in 1859, steady operation continuing for the first five months of 1860. During June and July of that year the glassworks again was being rebuilt, but production was resumed in August. At this period agriculture was also stressed. Cranberry cultivation is mentioned in the records for the first time, in such entries as these:

1860:	September	5—Five men at cranberry fields.
	October	2—Sowing rye. Cutting hay.
	November	21—Planting apple trees.
1861:	May	10—Planting potatoes.
	May	24—Hay from Harrisville.

These developments were not particularly significant. The glass factory remained in operation most of 1861, with shipments going out by rail, via the Atlantic City Railroad at Hammonton, and by the schooner *Rebecca*, but an ominous notation marks the close of 1861: "In general business seems to be falling off."

At this point it is necessary to go back a few years, to 1846. On December 21 of that year a young man named Robert Stewart came to work for Jesse Richards as a bookkeeper. His salary was to be $600. He was to have a house rent-free and a 20 per cent discount on goods from the company store. Stewart soon became the most trusted employee at Batsto. More and more responsibility fell upon him, especially after Jesse Richards' death, and it is to him that credit is due for most of the records kept in the Batsto Store Books.

Robert Stewart's destiny was to preside over the industrial collapse of Batsto: its commercial disintegration, its financial bankruptcy, and its physical destruction, in large part, by a devastating fire. In October of 1860, weary, no doubt, of increasingly uncongenial tasks, Jesse Richards' heirs turned over to Stewart the management of Batsto. Thomas H. Richards continued to live in the Big House, but during the remaining years covered by the Store Books the name of Richards appears less and less. Stewart "commenced on his own account" as manager at a time when business was in the doldrums. The glassworks was running, but another fire halted operations for nearly three months in 1861. Due perhaps to Civil War prosperity there was an upturn in 1862 and 1863 when the glass factory worked almost full time, despite a "thirteenth glass house fire" in June, 1863. A "fourteenth fire" occurred in May of 1866, but even then the factory was rebuilt.

The next year, 1867, was a grim one. On January 5 the "batch room" burnt down. On February 12 occurred an event which showed that, besides its other woes, Batsto was struggling with the pressures upon its feudal way of life from the "free enterprise" economics characteristic of those times. The significant event occurred when the "boys refused to work on account of not receiving cash." This was the first hint of anything approaching a strike at Batsto. Presumably some cash was found, but very shortly the "blower boys" again "struck for cash." This strike appears to have lasted until the 25th, when a note appears: "Started again."

It was during this period that the loss of large sums through the New York agency plunged the Richards estate still deeper in financial difficulty. To tide themselves over, the heirs sold 30,000 acres of their landed estate to the Batsto Farm & Agricultural Company for a real estate development. Later they sold another 25,000 acres.[29] One mortgage had been running on the Batsto works since 1845, and new debts provided no answer to old ones. More wages went unpaid, and evidently they stayed unpaid. These included the wages of Stewart himself, and the courts granted him a judgment for some $20,000. Other creditors pressed their claims, and by 1868 Batsto was in receivership, its enterprises at a standstill. Hopes persisted that those enterprises could be revived. Soon, however, those hopes faded. Members of the Richards family no longer came and went at the Big House. It was far from Batsto that one of the clan later was to carry on its "iron tradition," John Richards being an executive of the Bethlehem Steel Corporation for forty years, until his death in 1955. Before long the mansion was empty, and falling into disrepair. Many of the workmen moved away. Schooners long since had ceased coming up the Mullica for Batsto products. Farmers began taking their goods to other markets. No capital was in sight for new ventures. There was not even enough money to maintain the plantation.

One of the most pathetic figures in the debacle at Batsto was Robert Stewart, who had served not only wisely, but too well. A letter of the 1860's sent to a relative by his wife, Margaret, tells a bitter tale. She writes:

> In the first place we let Mr. Richards [Jesse, Sr.] have $500 in gold at 6%. Robert's salary was $600 at first and then $700 for a longer time to the war when it was raised to $1200.
>
> Andy went into the store when he was 11 years old and Rob to cut glass when he was 13. Also William. Jennie to teach school when she was 15 years old. We also raised a number of cattle every year which were sold to Mr. Richards but never paid for. Also we got some money from home [Northern Ireland] and that was used to maintain the family. All our earnings the Richards got . . . and at the end left us without one cent. . . .
>
> After the work was stopped entirely Robert was offered situations elsewhere but the Mr. Richards would not hear of him going. They gave him four promises [promissory notes, apparently]

and have left us here quite destitute while they can live in luxury and keep a houseful of servants, and all the rum they want to drink.

So now, after the labor of a steady, industrious family for 22 years, here we are, without one foot of ground or one dollar.

Margaret Stewart's mention of "all the rum they want to drink" raises a question as to how great a part liquor played in the decline and fall of Batsto. Some views on this are to be found in the *New Republic*, a Camden weekly, issue of April 18, 1874:

Jesse R. [Jesse Richards, Jr.] walked around the ruins of his native place on Friday, and said nothing, drank at the tavern in Pleasant Mills, and went to his residence in Lower Bank. "If he had but let whiskey alone," said his old manager on Saturday, "he might have been up there in the Big House and the first man of Batsto." The Postmaster at Pleasant Mills, pointing to the visible chimneys standing like sentinels over the desolation, said: "Whiskey has killed that place." And pointing to an inebriate curled up in the sun against the Post Office, "and it is killing its people."

The "ruins" mentioned in the newspaper article were the result of a great fire which swept Batsto on the night of February 23, 1874. Manager Stewart's house was set ablaze by sparks from his own chimney. Soon a neighboring house was burning. Bucket brigades proved futile, and there were no other facilities for fighting a major blaze. The fire continued to spread. Dwelling after dwelling was destroyed. By morning 17 of Batsto's little houses were in ashes. Gone were the husks of the glass factories, gone, also, the last buildings connected with the old furnace. The Big House survived. So did the gristmill, the barns, the store, and one venerable house on the south shore of the lake, where eight of its neighbors lay in a long row of ruins. Saved, by some good fortune, was that portion of the village now standing; but half the families of Batsto had lost their living quarters, and most of their meager possessions.

Gone, too, was any lingering hope of staving off financial disaster. In 1876 foreclosure proceedings were completed on the $13,425 mortgage of Jesse Richards dating back to 1845. This took precedence over other claims and judgments. Batsto was put up at a Masters Sale. It brought only $14,000. Its purchaser was Joseph Wharton.

Batsto seems to have fascinated Joseph Wharton as it had so many before him. Its purchase was one of his earlier pineland acquisitions, and he had great plans for it. How much he spent on rehabilitating Batsto is uncertain, but he is said to have expended more than $40,000 on the mansion alone. He chose as his manager General Elias Wright, distinguished veteran of the Civil War. Wright supervised the repairing of the buildings and also handled most of Wharton's subsequent real estate transactions. He writes that when Wharton got title to Batsto, "The buildings were so dilapidated that there was much doubt as to whether they should be repaired or torn down."

Wharton first purchased pineland property to experiment in the raising of sugar beets. His big dream, of course, was to turn the area into a source of drinking water for Philadelphia. After that project came to nothing Wharton's interests were largely agricultural. Great stress was placed on cranberry culture, which has prospered to this day. About 1881 Wharton took up livestock breeding. One record states that he started with a herd of nine cows, at Batsto, and that further acquisitions soon increased their number. A breeding farm was established five miles away at Washington, and the ruins of the big stone barn there are still visible. As of January 11, 1884, Wharton's total livestock is listed as follows:

At Batsto (calves and all)	113
Washington	50
Hogs at Batsto	17

Wharton is believed to have spent a fair amount of time at Batsto, in the Big House which he had remodelled extensively. It is impossible today to trace with very much accuracy the history of the various additions, alterations, and "improvements" to that mansion. The section to the rear is itself a "complete house," the earliest portion of the standing structure, probably erected before 1800. It is known that from Charles Read's time there was an "ironmaster's dwelling" on the site, and the house of "W. Richards" is marked on old maps. It is not certain who built the vast

"Victorian" portion of the present mansion and set it in front of the older part like an elephant overshadowing a gazelle. Probably Jesse Richards was responsible, since it dates from well before the Civil War. Pre-Wharton photographs show the front wall of the house much as it is now. The chief Wharton alteration there was the substitution of one large window for two small ones on the third floor. Wharton, however, built the large northern wing, the tower, and reconstructed the massive staircase. Bertram Lippincott, a member of the Wharton family who knows the house well, has this to say:

> During the process [of rebuilding] the entire inside and outside of the house must have been plastered and new woodwork put in, as it is now the same throughout, making it very difficult to "date" the original building from an architectural standpoint. Some foundations in the cellar, however, and possibly some of the present walls may be of the original house.[30]

It has been suggested, with some plausibility, that the earliest ironmaster's dwelling may have been the western, three-story portion of the "company store." This store, like the mansion, gained an addition in later years. The type of construction lends some support to this theory.

In the Big House there are 36 rooms of all sizes and shapes. The most glamorous of these is a "secret" hiding place, said to have accommodated fugitives for the underground railroad during slavery days. This low-ceilinged room is above the third floor and is approached through what once was a false back in a closet. A few short steps beyond the closet lead to this room, which overlooks the front of the estate through a semicircular window. The most dramatic showpiece of the Big House is the tower, and a trip to the top repays the effort of climbing the winding staircase. This tower, 86 feet high, contains a water tank and is surmounted by a glassed-in observatory which commands a view of the countryside in all directions. For some years Batsto's caretakers have lived in the "original wing" of the house, to the rear. This is in many ways the most comfortable portion. In its kitchen may still be seen a row of bells—each with a distinctive tone—all connected by long wires to the various bedrooms. The old-fashioned bellpulls in those bedrooms are still in working order, symbols of gracious living in an earlier day.

Exploring Batsto can be a rich and rewarding experience. Down the hill from the mansion stands the company store. Here is kept an array of interesting relics. Prominent among them is the old countinghouse desk used in the post office, its sorting rack above, and inside the desk's lid is a sticker with mailing instructions for days when stagecoaches were carrying the posts. Here, too, is an early Batsto stove; the bellows from the old blacksmith shop; and an array of tools and oddments—including a weather-beaten feminine figurehead from the prow of an old ship. At the end of the building is a vault, with combination lock. This was once the scene of an attempted robbery. One of the most interesting exhibits of all, a gradually stepped flagstone walk, rises toward the Big House from the easterly end of the store. These flagstones are slabs of iron—enduring bog iron, almost impervious to rust.

Across the road is the gristmill, a yellowish-plastered stone struc-

BATSTO FURNACE

KEY:
  1 The "Spy House"
  2 Sawmill
  3 Site of Glass Furnace
  4 Site of Batsto Furnace
  5 Batsto Mansion
  6 Store and Post Office
  7 Barns
  8 Gristmill and Corncrib
  9 Old Stone Barn
 10 Blacksmith Shop
 11 Water Tower
 12 Stable
 13 Probable Site of Early Dam
 14 Site of Furnace Stack; later a "Carp Pool"
 15 Present Dam
 16 Trace of Old Canal
 17 Reputed Site of Cupola

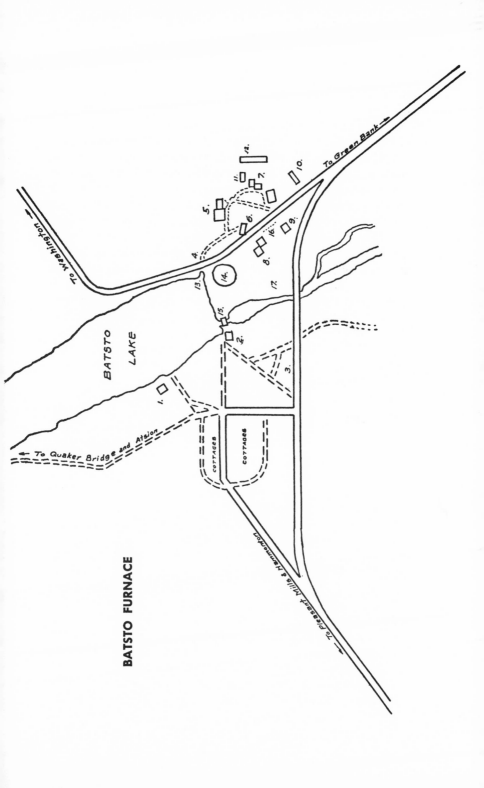

BATSTO FURNACE

ture, which bears a plaque: "1828." It was built by Jesse Richards and is still in operating condition, with its two great wheels intact. One of these is dressed for milling wheat, the other reputedly for milling corn. The adjacent corncrib is mounted upon iron posts cast at Batsto Furnace. Other surviving buildings are the two stone barns, with long ventilating slits in the walls. One barn is dated "1830"; the other is probably some years older. Finally, beyond the grove of towering buttonwoods, there are the stables, built by Wharton, with one more eye-catching tower. This was a water tower, not a shot tower as another legend would have it.

Beyond the dam, over which the Batsto River spills and tumbles, come the sawmill, and a road which leads to the village itself. The present sawmill, still in full operating condition, was built by Wharton in 1882, and a contemporary newspaper account mentions that "It has five saws and a greater capacity than any mill in this vicinity."

A stroll through the village conveys the impression of refreshing serenity and restfulness. Here are 24 cedar cottages arranged on three streets. These dwellings have long been leased to the occupants, as they were during the Richards era. A lease form of Joseph Wharton shows that rentals in his day were $2 a month, provided that the lessee agreed "not to sell or otherwise deal in spirituous or malt liquors or tobacco, or to fish in the stream or pond above the Batsto dam. . . ." One recent lessee has a sense of humor. Nailed to a tree in front of his house, at the corner of the road through the woods to Atsion, is a sign: "Broadway & 42nd St."

To the left, along the lake bank, is the oldest house in Batsto. Almost certainly it was built in colonial times. This was one of the nine houses along that shore, the one that escaped the great fire. Mr. and Mrs. Charles Simpson, who have occupied it for some years, state that it is commonly called the "Spy House." Legend has it that the place, which is just off the road to Atsion and Quaker Bridge, was used by a Tory during the Revolution to keep check on the running of munitions to Philadelphia. Below the Batsto River, too, near the main road to Hammonton, traces may still be seen of the foundations of one old glass factory. These are on the left side of the highway, just below the bridge.

Beyond Batsto proper, inviting journeys await the imaginative hiker and researcher. To the east is that pleasant bend in the

Mullica River where "A. Nichols" once had his store. There elections were held and ships came in bearing ore and other necessities for the Batsto plantation. Properly equipped to cope with brambles, and with a willing guide, an explorer may visit the old Batsto Forge Pond, about half a mile southwest of the village. Here, by the Nescochague River, remains of an old dam may be seen, and it is possible, even today, to find the old slag pile, deep in the forest tangle. A five-mile jaunt to the north—along what has been euphemistically called the "Batsto-Washington Turnpike"—brings the traveller to the remains of the old cattle-breeding farm of Joseph Wharton at Washington. Not far from Batsto itself, in the Pleasant Mills burial ground, lie many who helped smelt the iron, make the glass, saw the wood, till the soil, and tend the households during Batsto's fruitful years.

Batsto's next chapter is about to be written, under the auspices of the State of New Jersey. With the old mansion opened to visitors, at specified hours, and other restoration projects in the making, the village may well become a miniature Williamsburg in the Pine Barrens. Few who have visited Batsto have failed to succumb to the loveliness of its natural setting and the fascination of its history. Its appeal is more than sentimental. The United States today is the world's most successful democracy and its greatest industrial complex. The men of Batsto made substantial contribution to both. May the people of New Jersey, the new owners of Batsto, always keep the memories of those men as green and fresh as their woodland background.

10
ETNA FURNACE AND
MEDFORD LAKES

Many years ago the Indians of New Jersey followed the Shamong Trail from the site of Burlington southeastward through Atsion to the seacoast. Six miles from Burlington the trail touched Mount Holly. Seven miles farther along it crossed Bally Bridge, which was to be renamed first Upper Evesham and later Medford. About three miles below Medford the Shamong Trail crossed a small stream, which still flows quietly today beneath a small, white, wooden bridge.

This little bridge is at present the threshold of a rustic community called Medford Lakes. Nearly two hundred years ago the bridge adjoined the site of an ironworks and town known as Etna Furnace. For ages before that, this particular spot had been a sort of entrance into the New Jersey Pine Barrens. At the bridge the terrain and atmosphere change completely. To the north is typical South Jersey farmland. Across the stream, however, the soil suddenly ceases to be fertile and instead is sandy and often as white as a seashore, which it used, indeed, to be. Here open fields yield to great forests of pitch pine and scrub oak, tangled with laurel and holly, often giving place to extensive bogs and strange swamps of the white cedar. Even the character of the streams themselves is different, for until they meet the tides, Pine Barrens rivers are of dark, soft "cedar water." Many have found, like the author, that this quiet pineland offers an atmosphere of contentment and calm.

Long before Etna Furnace was built, tradition has it, the Indians were accustomed to foregather in this location. There is a great tree still standing, on a slight eminence overlooking two lakes, which is supposed to have been a landmark and rallying point for the red men. At any rate, their presence has been established over the years through the finding of arrowheads and other Indian relics in the vicinity, and for that reason the modern community has given Indian names to most of its lakes and trails.

The first white men to establish themselves in the vicinity were the operators of sawmills—the merciless massacre of Jersey pine was a major industry for years—but there was nothing resembling a real community until 1767. At that time Medford was beginning to be a village, with a schoolhouse "near Robert Braddocks" which was used for the Friends' meetings.[1] What's more, the Shamong Trail had become a fairly well-travelled road—narrow, sandy, treacherous, often soggy and sometimes impassable—but still one link between Burlington on the Delaware and Clamtown, later Tuckerton, which was then a busy port on the Atlantic coast.

In 1766 Charles Read, of whom much already has been said, began building Etna Furnace and the community around it. The little stream already mentioned was but one of several which joined nearby to become the south branch of the Rancocas River, and Read was quick to recognize the value of these streams for water-power. Read also saw that the location possessed three other essentials for the production of iron: supplies of bog-iron ore close by; abundance of woodland for the making of charcoal; and access to cheap transportation, which in this case was the once-busy landing at Lumberton, below Mount Holly.

Charles Read bought the site of Etna Furnace from John Prickett, John Budd, and Levi Shinn.[2] He acquired additional tracts in the vicinity from Enoch Haines, Caleb Haines, and "Carlile" Haines, and had other lands surveyed to him from the West Jersey Proprietors. By the time Etna was in full swing Read had 9,000 acres, together with "ore rights" to still other nearby lands. These "ore rights" he obtained through an agreement with David Oliphant whose forebears in 1685 had built "Oliphant's Mill," a gristmill near the new St. Peter's Episcopal Church. Before their purchase by Read the Etna lands had been known as the "Saw Mill Tract," for a "swift-going saw mill" which Read leased for 200 pounds a year. The name "Saw Mill Tract" was later revived, after the furnace fires died out, and the sawmill itself, or successors on the same waterpower, prospered well into the twentieth century. What had been the Shamong Trail and is now prosaically called "Stokes Road" was in Read's time and later known as "Mountain Run Road." It was named after a stream—Mountain Run—which feeds the lake named Ockanickon, and flows at the base of a considerable hill about a mile below Etna.

The best description of old Etna Furnace was given by Charles Read himself in an advertisement in *The Pennsylvania Journal* of October 11, 1770, offering the property for sale "on reasonable terms." The advertisement read, in part:

> TO BE SOLD—ETNA FURNACE—In Burlington County, 19 miles from Philadelphia, situated on a tract of near 9000 acres of well wooded land, in a plentiful and healthy country, furnished with all houses and buildings necessary . . . 7 miles from the present landing on Ancocas Creek, & one may be made at four miles distance, whence boards are daily floated in rafts of 2500 feet. The land carriage to New York market is 17 miles. To-

ETNA FURNACE   (*Now Medford Lakes*)

KEY:

1 Etna Furnace
2 Lower Sawmill
3 Springhouse
4 Blacksmith Shop
5 Old Frame Cottage
6 Gristmill
7 Company Store (*brick-floored*)
8 Charles Read House
9 Icehouse
10 Cranberry Pickers' Shop (*probably of much later construction*)

NOTE: Except the Charles Read House, still standing, these are site locations only, as the buildings have long since gone. Mountain Run Road is now called Stokes Road.

ETNA FURNACE

gether with a small stamping mill and a grist mill close to the
furnace, and conveniences therein to grind and polish iron ware
by water; there is in the furnace a variety of nice patterns and
flasks for casting ware, for which as well as barr iron the metal of
this work is very fit, and is in high esteem at foreign markets.
. . . There is plenty of ore at different distances, and the roads
so remarkably good that the carriages will last many years with
very little repair.

Etna Furnace itself almost certainly was located along the base
of the elevation near the foot of the present Lower Aetna Lake
(Aetna has been spelled both with and without the A, but was
usually spelled without it in Read's time). It was desirable in those
days to find just such a spot for the building of a furnace, that is,
a hillside adjacent to a source of waterpower. This permitted the
placing of the water wheels and furnace machinery on the lower
level, while on the upper level would be the "furnace bank" with
its charcoal house, weighing station, and piles of ore and flux.
The present topography suggests that the top of the furnace here
was about level with the top of the hill, which not only made for
an approximately level trestle from the "furnace bank," but took
full advantage of the fact that Mountain Run Road had an easy
grade which reduced the uphill hauling of furnace materials to a
minimum. The accompanying map shows the probable location of
the furnace and other buildings so far as can be surmised today.
The Aetna Lakes, of course, are named for the old furnace, and
there was in Read's time a smaller pond across the road which is
today called Ballinger Lake.

As will be seen further from the sketch map, the famous old
sawmill was located just across the dam from the furnace, near the
present bathing beach. A stream still follows the course of the old
tailrace from the furnace, which carried off water the force of
which had been spent by the mill wheel. Read also established a
gristmill as one of the community enterprises. This was near the
present dam of Ballinger Lake. Close by was the company store.

In connection with Etna Furnace Read built a three-fire forge.[3]
The works thus were equipped to produce not only bar iron—
malleable iron refined from pig iron—but also ironware in the form
of flatirons, kettles, iron-handles, firebacks, and other articles in
commercial demand. Some of the patterns used at Etna displayed

a degree of artistry, as is suggested by the two known surviving firebacks. One of these is decorated with a stag, the other with a cock crowing.

Of the various buildings erected in the time of Etna Furnace, only one still stands, in the author's opinion. This is the house at the left of the road just after the entrance to Medford Lakes from the north. Here Read spent his last few years, and here his son, Charles Read, Jr., who managed Etna Furnace, lived until his own death in 1783. This old house, with its brick-lined outer walls and solid brick inner walls, has undergone changes over the years. Part of it probably was rebuilt in 1773, as will be indicated later, and there have been minor additions and inner remodelling. Basically, however, it is Charles Read's house, one of the two surviving dwellings in which the elder Read is known to have lived, and the only "ironmaster's house" which he ever occupied in any of his iron enterprises.

Read's biggest problems in the operation of Etna Furnace concerned labor and liquor. Iron production required skilled as well as unskilled labor. The former was hard to recruit, and the latter hard to retain. The following advertisements give some idea of Read's difficulties:

From *The Pennsylvania Journal*, March 3, 1768, comes this appeal:

> At Etna Furnace, in the County of Burlington, good Colliers, two good Carpenters, a good Smith that understands the making of flatt or padd iron handles, a Stone Cutter, a person used to grind flatt irons and wagon boxes, will meet encouragement.

In November of the same year the *New York Gazette* carried the following notice:

> ETNA FURNACE. . . . Wanted at said furnace, a good Keeper or person accomplished in Castings, and a middle-aged Woman not subject to Liquors; fit to be entrusted with the Care of a large Family but not to cook.

Early in 1769 another advertisement in the same publication read as follows:

At ETNA Furnace, Burlington County, which will begin to be in blast the Middle of April, are wanted, a good Keeper, two Master-Colliers, Moulders and Stock-takers: Those who apply in season, and come well recommended, will be encouraged. The Wood should be running in February, or beginning of March.

Read appears to have had his greatest difficulties with indentured servants, upon whom he seems to have relied for most of his semiskilled labor, while his Negro slaves, on the other hand, gave him comparatively little trouble. Three of the former were sought through the following advertisement in the *Pennsylvania Chronicle* of June 4–11, 1770:

THIRTY DOLLARS REWARD
RUNAWAY from Etna Furnace, last Night, three Servant Men, the one a German, who has been long in the country, named PHILIP NOTT, about 5 feet 6 inches high, square sett, swarthy Complexion, black hair, a Hair Mould on his cheek, sometimes pretends to walk lame; about 35 years old.

Another named PHILIP JONES, a Welchman, aged about 24 years, fair Complexion, thin visage, speaks hoarse, grey eyes, and lightish Hair; about 5 feet 6 inches high.

Another named ALEXANDER CAMPBELL, aged about 22 years, sharp, thin visage, small Eyes, fair Complexion, lightish-coloured hair, much given to Liquor, and chews tobacco, in his Liquor noisy and quarrelsome; has worked at a Forge Fire; He writes a good hand, and may forge Passes. Whoever secures any of the said Servants, so that they may be had again shall receive SIX POUNDS for the First, THREE POUNDS for the Second, and TWO POUNDS for the Third, paid by
CHARLES READ

A more detailed description of the type of workman to be seen around Etna Furnace in those days is given in this advertisement in the *Pennsylvania Gazette* for July 4, 1771:

RUN AWAY from Etna Furnace, in the Jerseys, the 22nd of June last, from the subscribers, two servant lads, one named WILLIAM HOPKINS, aged about 20 years, about 5 feet 6 or 7 inches high, of a dark complexion, and has a down look; had on an iron collar, a check shirt, a pair of tarred trowsers, a brown double-breasted jacket lined with homespun stripe, with brass buttons, an underjacket, without sleeves, much darned on the back, a new ozenbrigs frock, without buttons or holes, had neither hat nor shoes. The other, named JOHN, has lost the sight of one eye, aged about 21 years, about 5 feet 4 or 5 inches high, of a light complexion, pretty talkative; his apparel is not known, for as it is thought he has changed his clothes. Whoever takes up the said servants and secures them in any of his Majesty's gaols, so that their masters may have them again, shall have FIVE POUNDS, and all reasonable charges paid by

CHARLES READ
WILLIAM EVENS

Liquor was a serious and complicating factor in labor problems of those days. Largely due to Read's powerful political influence the New Jersey Assembly in 1769 enacted a law giving owners of ironworks the right to furnish employees with "Rum or other strong liquor, in such Quantity as they shall from Experience find necessary." The same statute provided that, "If any other Person or Persons, residing within Four Miles of said Works . . . should entertain in or about their House, in idling or drinking, or shall sell any strong drink, to any Wood-cutter, Collier or Workman employed at said Works," the offender was subject to a fine of ten shillings for every offense. Money from the fines was to go for road maintenance.[4]

The effect of this legislation was to prohibit the operation of taverns within four miles of any of Read's iron establishments. The same law gave other encouragement to ironworks development by exempting from taxation the Etna company store and similar stores at other forges and furnaces, the tax exemption being extended to the gristmills if they were used "for the sole purpose of supplying the Persons and Teams necessary at said Works, and not taking toll, or manufacturing Flour for Exportation." Broadly speaking, this statute gave State approval to the rustic feudalism which soon developed not only at Etna, but at most of the other industrial enterprises in the Pine Barrens.

The mingling of mystery and tragedy which marked the final destiny of Charles Read descended in full measure to his heirs, and to Etna Furnace itself. Three sons had been born to Alice and

Charles Read: James, who died in infancy; Jacob, a wastrel and drunkard; and Charles, Jr., born in 1739, the manager of Etna Furnace.

Charles Read, Jr., was assured possession of the furnace through the deed executed by his father in April, 1773. Yet as the elder Read wrote to Israel Pemberton, concerning his iron business: "My son disliked it." Whether that dislike led to the closing down of Etna Furnace is not yet certain. It is clear, however, that Etna Furnace did not long survive the hasty departure of Charles Read, Sr., from New Jersey. Boyer states that Etna "subsequently passed to William Richards of Batsto." This is an error, as a tracing of the title now shows. All evidence indicates that the fires of Etna Furnace died down, and its hammers were stilled, in that same fateful year of 1773.

That Etna Furnace was operating in April of 1773 is certain, for Elizabeth Drinker, in her Journal for that month, tells of her trip to Atsion, where her husband was a partner. She writes: "We stopped at Charles Read's ironworks, 10 miles from Moorestown." Subsequent letters of the younger Read, up to September of that year, suggest nothing untoward. On November 25, 1773, however, writing to Israel Pemberton, Charles Read, Jr., not only mentions financial difficulties as preventing his payment of a debt and asks a further loan of "about 100 or 150 pounds," but he also speaks of "having every Necessary for Housekeeping to buy," which "pushes me hard to save my credit." He further mentions that "I am now putting my place in good Order and am in hopes by Spring to have a tolerable show of meadow." [5]

What had happened? A fire? Had the Etna works been destroyed by uncontrolled bursts of flames from their own furnace, a not uncommon occurrence in Pine Barrens iron towns? Or was Etna—both the ironworks and Read's own home—in the path of one of the forest fires which seem to have been as prevalent and devastating in those years as they are today? At any rate, the following facts are definite, and indicate that the furnace itself was no longer in operation:

The younger Read had to re-equip completely the house he had occupied for at least five years.

He became a farmer, in 1773, after having been an ironmaster since 1767.

The following year—1774—he advertised in the *Pennsylvania Gazette*, offering to sell "Two Forge Negroes, One a Good Finer, and the other a good Hammer-man." With such labor scarce, Read hardly would have disposed of those slaves if the furnace were still in operation.

And Charles Read, Sr., records now show, sold the Etna furnace machinery to Henry Drinker and Abel James, when he sold them his interest in Atsion.

Since Taunton, Atsion, and Batsto all made munitions during the Revolution, there probably would be a record of similar activity at Etna, if Etna had been operating, even though Charles Read, Jr., later turned Tory. No such record ever has been found. This explains why.

Thus the Etna of Charles Read, Jr., was an agricultural community. But still other mysteries have surrounded the son of its founder. Charles Read, Jr., in 1767, married Ann Branin, the daughter of Michael and Elizabeth Branin "of Burlington County." So much historical confusion has been woven about this luckless couple that a few facts concerning them are in order here, and perhaps the best way to offer those facts is to contrast the truth with the legends.

First, Charles Read, Jr., has often been confused with his father. For example, a South Jersey history states at one point that the elder Read was alive in 1776 and a few paragraphs below states that he died in 1774. The first reference, of course, is really to the son.

Second, for a hundred years Charles Read, Jr., was confused by historians with Adjutant General Joseph Reed, of General Washington's staff, and Joseph Reed's loyalty was long beclouded as a result of Read's defection to the British during the Revolution, of which more will be said.

Third, it has been stated in the Clement Papers that Charles Read, Jr., in his later years moved "to a small place between Medford and the Cross-Roads." This is false, for his correspondence is inscribed as coming from Etna right up to 1783, the year of his death.

Fourth, also in the Clement Papers there is mention of a report that "Charles Read" was "rocked in a cradle like a child for several years before his death." This could not refer to Charles Read, Sr., and it is wholly untrue with respect to Charles Read, Jr., for his letters, in the year of his death, show a sound mind beyond question. This is an old wives' tale and probably has its origin in the "itch of the Country people for tattling and reports without Foundation" which the elder Read mentioned.

Fifth, there is a legend that Charles Read, Jr., was married twice, and had two sons named Charles and two daughters named Alice, one set by each wife. This, again, is false. The Reads did have two sons named Charles and two daughters named Alice, but all were the children of Ann Branin. The first Charles died December 6, 1769, at 13 months; and the first Alice died three years later. An obvious desire to carry on these family names accounts for giving them to subsequent offspring.

With these confusions disposed of, it is possible to return to the career of Charles Read, Jr., at Etna. While his father and grandfather had been Episcopalians, his great-grandfather had been a Quaker, his Philadelphia kinsmen—the Logans and Pembertons—were Quakers, and so it is not surprising that Charles Read, Jr., embraced the Quaker teaching. Despite this, he had in 1764 accepted a commission as lieutenant of the Burlington County militia, and held the rank of Colonel when the Revolution broke out. In 1776 he was appointed deputy to the New Jersey Convention to draft a new State Constitution, and that same year found him in command of militia from various South Jersey counties. Thus he seemed in a fair way to redeem the family name, and to establish himself on his own account in the state which his father had served so long.

That, however, was not to be. In his *Ploughs and Politicks*, Carl R. Woodward tells how during November, 1776, Charles Read, Jr., was put in charge of a battalion known as the "Flying Camp" of New Jersey. One month later he went over to the enemy. The British had offered "protection" to all who would lay down their arms. Read accepted the offer, abandoned his command, and retired to his home at Etna. Shortly afterward, while travelling between Mount Holly and Moorestown, Read was captured by Colonel Samuel Griffin of the Continental Army. Read is said to have told Griffin that "He was not disposed to serve any longer," whereupon Griffin called him a "damned rascal," and took him with other Tory prisoners to Philadelphia. There he was placed in custody. In the same year his kinsman, Israel Pemberton, was seized and taken to Virginia for imprisonment.

Read describes his action, in a letter of April, 1783, as "the Disagreeable Necessity of taking Protection from the British." [6] It was this act which for over a century was blamed upon Adjutant General Joseph Reed, an historical wrong which was not righted until William S. Stryker wrote his *The Reed Controversy* in 1876.

Read was released from prison on January 21, 1777, on his promise not to leave Philadelphia without permission. By December of that year, however, permission apparently had been given, for on December 17, 1777, he mentions carrying some rye to Oliphant's Mill, where "They took no toll and I had 765 lb Meal which is 38¼ lb Meal to ye Bushell." Obviously Read was then back at Etna.

Some idea of Read's household at Etna can be gained from the inventory of his personal property.[7] In the Read living room were a walnut table, six chairs, a mahogany desk, a bookcase, and a corner cupboard. There were a clock, six pairs of candlesticks appropriately placed, a mirror on the wall, and andirons, shovel, and tongs at the fireplace. In the dining room the Reads had, besides the dining table and chairs, a tea-table, "Chinea" ware, "sundry tumblers and glasses," and the usual fireplace equipment. Upstairs, the master bedroom contained a "bedstead and bedding with a red & white coverlid on," a "Double Case of Walnut Drawes," bureau table, cradle, "trunnell bed," walnut armchair, and large "looking and shaving glass," with fireplace equipment and sundries. Apparently other bedrooms were furnished, and there was additional emergency bedding in the "Loc Garratt."

Out in the barn were to be found a sleigh, a mule cart, a sulky and harness, a "good four-horse waggon," a "new light waggon & swingle-trees," and "geers for the team." There were the usual farm implements, a "Barrell Churn," a cheese press, and a hog trough. The livestock included four team horses, two carriage horses, a mule, seven milch cows, three sheep, ten geese, three sows and "11 pigs at the house," a pair of oxen, and two young heifers.

Mention is also made of a blacksmith shop, sundries in the gristmill, and two sawmills—the "Lower Saw Mill," with a "broken crank, 5 saw mill saws, 3 Crow Barrs and a Cross Cut saw," and the "Upper Saw Mill," where there were at the time 4,000 feet of pine boards. The "Lower Saw Mill," of course, was the one opposite the furnace site.

The Read cradle and trundle bed were put to good use; in their 16 years of married life the Reads had nine children. Four of these died at an early age, the first Charles and Alice, already mentioned, who are buried in the Friends Cemetery at Mount Laurel, and also Samuel and Ann, the grave of the former being in the Friends Cemetery at Medford. Rounding out the Read household, at least in its later days, were three slaves, Rachel, who was set free in Read's will, and two boys, Richard and Phil, whom he directed to be bound out and to be freed at the age of 21.

Charles Read, Jr., passed away at 44, a comparatively young man. Whether his humiliation during the Revolution affected his health can only be conjectured, but it almost certainly affected his social status at a time when a new nation was being born; he had staked his reputation on a cause which went down to defeat. His death appears to have been somewhat sudden. A letter he wrote earlier in the year contains no suggestion of ill-health, but on November 20, 1783, Read died, leaving his wife and five children—Charles (fifth of that name), William Logan, James, Alice, and Elizabeth. To them he bequeathed his estate. There seems to have been little public notice of his passing, and even the whereabouts of his grave—like that of his father—is unknown.

One more tragedy awaited the family of Charles Read. Five years after his father's death, the fifth Charles became involved in a drunken brawl in Philadelphia, where he appears to have been employed as a hatter. In the brawl he drew a knife "of the value of six pence" and stabbed to death a boatman named Andrew Homan. In January, 1789, he was tried for murder, convicted, and hanged.

Etna—soon to be Etna Mills—passed out of the Read family on May 1, 1784, less than six months after the death of Charles Read, Jr. On that day the Etna property—only one of many which Read had owned—was conveyed by Ann Read and Samuel Allinson, a prominent attorney, as Read's executors, to four men: Joshua Bispham, Caleb Austin, Jabez Buzby, and Amos Bullock.

The records show that Bullock purchased Bispham's quarter

interest in Etna on June 6, 1785, after which the three remaining owners divided the property into two parcels by exchanging deeds of release. In this exchange "Parcel A," which included the old Read house, the gristmill, the Lower Saw Mill, and about 61 acres, went to Amos Bullock. "Parcel B," the balance of the tract, included the Upper Saw Mill and about 8,000 acres of land. This went half to Buzby and Austin, half to Bullock.[8]

Various transactions followed by which the tracts were conveyed in part to William Sharp, and in part to Jonathan Crispin. Sharp apparently controlled "Parcel A" with the gristmill, and Crispin "Parcel B." Sharp negotiated the sale of his interest in 1792, and again in 1794, but on both occasions the transactions backfired when the purchasers defaulted on payment. Finally, on October 19, 1797, he disposed of the house and gristmill and the balance of his holding to Shinn Oliphant, who was to run the "Etna Mills" for over twenty years. Oliphant acquired the remainder of the original tract, including the sawmills, from Joseph Gardiner, on May 9, 1802,[9] Gardiner having bought Crispin's interest in June, 1795.

Thus Shinn Oliphant, in effect, restored the tract almost as it was sold by the Read estate.

Under Shinn Oliphant the two sawmills and the gristmill prospered. He lived in the old Read dwelling, and just across the road there still stood the onetime company store, with its characteristic brick flooring. Indications are, however, that it was not then used, save perhaps for grain bins. Most of the old ironworkers' houses were gone, and nothing was left of the furnace itself but fading memories. Soon even the spelling of the name was changed to "Aetna Mills," and that name, in its turn, was changed half a century later.

Many of the old Burlington County families intermarried over the years, and as a result, most of them eventually were related to each other. Thus it was not strange that the Oliphants were related to the Ballingers, and that when it came time to sell Aetna Mills the purchaser was Thomas E. Ballinger, who took title to the heart of the property on April 9, 1821. Aetna remained in the Ballinger family for more than a century, the most prosperous and at the same time the most quiet period of its existence.[10] Over those years the changes at Aetna were not great. New dams were built on occasion. An ice house was constructed across from the present Borough Garage. The Lower Saw Mill apparently was enlarged. Cranberry culture was tried out, and this experiment was so successful that in later years the upper portions of both Aetna and Ballinger Lakes were used for bogs. The old blacksmith shop remained in operation. An orchard thrived near the site of the author's home. In the main, however, the pace of living and working was measured, peaceful, and calm.

After Thomas Ballinger's death the title to "Aetna Mills" passed to Dudley and Eayre O. Ballinger, and a deed of 1858 mentions the Read house as the "premises whereon the said Dudley Ballinger now resides . . . commonly called Aetna Mills." [11] While some maps showed the location as "Ballinger's Mills," the name "Aetna" survived and is to be found on a ten-cent-due bill issued by Joseph E.O. Ballinger, dated "Aetna Mills, New Jersey, July 4, 1862." [12] Joseph E.O. Ballinger, with his brother Alfred, had acquired title to Aetna on March 23, 1858. Alfred Ballinger sold his interest to Joseph on December 28, 1859; the latter, who then had full ownership, lived in the old Read house where, indeed, he had been born. On January 12, 1865, Joseph leased "the grist mill together with horse stables, hog pen and tenant house with garden attached" to his brother Dudley for "$600 a year rent."

The son of Joseph E.O. Ballinger was Joseph W. Ballinger, and in later years the father ran the gristmill while the son managed the sawmill. After the elder Ballinger's death, the son inherited and managed the entire property. While most of the former Etna tract is now the modern cabin community of Medford Lakes, the old house remained in the Ballinger family, and long was occupied by Mrs. Helen Ballinger Henderson, the daughter of "Joseph W." Her historic home, until its destruction, was the only surviving link with the Etna Furnace community of Charles Read.

Long before old Etna became Medford Lakes, wayfarers had eyed the location with envy. Some attempted to purchase home sites from the late Joseph W. Ballinger. He always refused to sell, observing that he would part with all of his property or none of it. However, his widow, Laura F. Ballinger, decided to sell all but the homestead.

A Texas speculator, Captain Clyde W. Barbour, fancied the idea of developing a rustic summer colony on the site of old Etna. Surveys showed that within the Ballinger tract and adjacent available land there were no less than twenty lakes or potential lake sites. By that time—1927—much of the area had reverted to something like its primeval state. Outside the general vicinity of the homestead there was forest in all directions, broken only by bogs and the winding streams which fed them. Many of the old dams were out. A house on the site of the present Administration Building had become a haven for tramps. As a final measure of value— most of the tract was assessed at $3 an acre!

If the building of Etna Furnace is largely the story of one man, so the history of Medford Lakes is in large part the story of another—a real estate man, Leon Edgar Todd, who like Charles Read had a dream. Unlike Read, he found it possible to make his dream come true. Todd became Captain Barbour's representative. He scouted the area by plane, decided it could be developed into a log cabin summer colony, convinced Barbour, supervised the artistic laying out of roads, trails, lakes, beaches, parks, and building lots, and—in 1927—saw the first cabins erected. That same year Aetna and Ballinger Lakes were reclaimed, and the first "colonists" were organized into what is still the key organization of the community—the Medford Lakes Colony Club.[13]

From time immemorial it has been the fashion of real estate men to move into a community, promote, sell, and then go on to new ventures. Todd broke that tradition. He himself moved to Medford Lakes, remained active in every phase of its progress, and lived there until his death. The Todd home stands by the approximate location of the old Etna furnace. Todd was Mayor for 15 years, from 1939 to 1954, and served in many other capacities. Much of the success of Medford Lakes has been due to his ability to recruit the help of outstanding men and women in building a remarkable community spirit.

Medford Lakes started, of course, as a real estate development within the Township of Medford. It was planned around various lakes, and the first cabins built were those on Lower Aetna. Save for three previously established roads—Stokes Road (old Mountain Run Road), Tabernacle Road, and McKendimen Road—all the "streets" were purposely made winding, wooded trails without sidewalks, designed not for speeding cars, but for slow-paced traffic suited to the restful tempo of living in the pines. At first, Medford Lakes was thought of as a summer resort, with cabins and other facilities developed accordingly. Later, all that changed.

The first public building erected in Medford Lakes was a community gathering place called "The Pavilion." That was in 1928. Eleven years later, during a snowstorm, the Pavilion collapsed. Almost at once construction of the present Vaughan Community House, a fine auditorium, was begun on the same site. Completed in 1940, it was named for Charles Z. Vaughan, first president of the Medford Lakes Colony Club.

In 1929, Leon Todd bought out the interest of Captain Barbour in Medford Lakes, and that same year the old Tuckerton Road was developed along the rear of the property, while Upper Aetna Lake was transformed from a bog into a beauty spot. Medford Lakes Lodge, said to be the largest log cabin hotel in the United States, was built in 1930. It was designed by Joseph N. Hettel, Camden architect, and the interior decorations were planned by Helen Todd, whose enthusiasm for Medford Lakes kept pace with that of her husband. The two churches of Medford Lakes were built in 1931: the Cathedral of the Woods (Protestant) and St. Mary's (Catholic). Both are of the log cabin construction characteristic of the entire community.

A major public development occurred in 1936 when, with W.P.A. assistance, the $175,000 sewer system and sewage-disposal plant were built. This improvement testified to a gradual change in the character of Medford Lakes. It was becoming an all-year-round community. More and more houses were built for winter residence, so that at the close of its first ten years Medford Lakes was an expanding town of 238 cabins and a population exceeding a thousand. As of this writing—1957—the population has jumped to nearly 2,500, most of them permanent residents.

Thus transformed, Charles Read's Etna is one of the few ancient bog-iron furnace towns which stands today as a modern, thriving community. Taunton Furnace survives upon a smaller scale. Atsion and Batsto at least are shadows of their flourishing selves. Martha, Speedwell, Harrisville, Hampton, and many more have been all but wiped from the face of the earth. If there is little in the Medford Lakes of today to suggest the feudal Etna Furnace community of nearly two hundred years ago, at least the Read house still commands the village threshold, while the furnace is commemorated in the names of the two largest lakes. As for the pinelands themselves, their beauty, fragrance, and appeal has changed little over the centuries.

11

PLEASANT MILLS

History has often travelled the trails of the Pine Barrens. On October 22, 1778, a strange expedition followed the winding, sandy road to Pleasant Mills and The Forks of Little Egg Harbor River. Twelve wagons rumbled eastward through that corridor in the woods, all of them empty. Five days later this caravan returned. No longer were the wagons empty. Some were loaded with woolens and linens; some with glass and nails. Others bore foodstuffs, including loaf sugar and tea. One carried the sails of a schooner and half a dozen swivel guns. Two days afterward these wagons were unloaded at the house of Stephen Collins in Philadelphia. The goods were quickly disposed of, and half the proceeds went to a general of the American Army, Benedict Arnold.[1]

Arnold had sent the twelve wagons to "The Forks," and the occasion seems to have marked a major stride toward the dark treachery which was soon to follow. Arnold liked lavish living and usually was short of funds. While still at Valley Forge he had entered into what he considered a purely commercial arrangement through which he acquired a half-interest in the cargo of a schooner, the *Charming Nancy*. The *Charming Nancy* was then tied up in British-held Philadelphia, and Arnold's part in the scheme was to give Skipper Moore a permit which would get him through the blockade of privateers keeping a close watch on Delaware Bay. This pass was supposed to protect the *Charming Nancy* from "umbrage or molestation" by "officers and soldiers of the Continental Army."[2] It was not honored, however, by the captain of the *Xantippe*, a New Jersey privateer. He captured the *Nancy* and took her to Little Egg Harbor, later Tuckerton, which then was a major haven for the privateer fleet.

Arnold was furious. He felt better when a New Jersey judge of admiralty decided against the *Xantippe* and released the *Charming Nancy* and her cargo. New cause for consternation came when Arnold learned that a British fleet was on its way from New York with orders to wipe out the "nest of rebel pirates" at Little Egg Harbor. That British expedition comprised two frigates, brigs, schooners, sloops, three galleys—twenty sail in all. Bearing the commander, Captain Henry Collins, R.N., was the flagship *Zebra*. Leading the military arm of the expedition was Captain Patrick "Scotch" Ferguson, who later died in battle in the Carolinas. In the ensuing engagement the *Zebra* ran aground, and some of her barnacled remains were found by divers as recently as 1954. General Washington, meanwhile, had dispatched Count Casimir Pulaski and his Legion to aid in the defense of the Little Egg Harbor area. They travelled by way of "The Forks," paused briefly there, and unfortunately did not arrive at their destination until the day after Ferguson and his forces had attacked and destroyed the fort at Chestnut Neck. Before any shots were fired, however, many of the privateers in Little Egg Harbor had sought refuge far up the river—then the Little Egg Harbor and now the Mullica— with some reaching "The Forks." Among these, probably, was the *Charming Nancy*.

Arnold's worries over his "investment" increased when word came that the British were moving up the river, not only to pursue the privateers, but also to destroy Batsto Furnace. Arnold was desperate. Making use of a wartime Pennsylvania law, he commandeered the 12 wagons and sent them to "remove property which was in imminent danger of falling into the hands of the enemy." That Arnold himself possessed a half interest in that property was not mentioned. That the wagons were more urgently needed elsewhere did not deter him.

The British, fearing an attack by Pulaski's main army, turned back and gave up their thrust at "The Forks." Tradition has it that one of Ferguson's advance contingents was ambushed by local Patriot forces in woods along the road to Batsto. By that time, however, Arnold's wagons were safely loaded with the *Charming*

Nancy's cargo. Soon word got around. A scandal developed. After controversy, and, finally, a trial, Arnold was sentenced to a reprimand from his Commander-in-Chief. George Washington's rebuke was characteristic:

> The Commander-in-Chief would have been much happier in an occasion of bestowing commendations on an officer who has rendered such distinguished services to his country as Major General Arnold; but in the present case a sense of duty and a regard to candour oblige him to declare that he considers his conduct in the instance of the permit as peculiarly reprehensible, both in a civil and military view, and in the affair of the wagons as imprudent and improper.

Where *was* "The Forks"? The late Alfred M. Heston wrote as follows concerning the origin of the name:

> Near the headwaters of the Mullica River, about five miles from the village of Elwood, is a long, narrow island, midway of the stream and parallel with its course. The division thus formed is called "The Forks" and marks the head of navigation. In the time of the Revolution there were houses and barns on this island in which were stored many of the cargoes captured by the privateers.[3]

In actual usage, as letters of the period show, "The Forks" was a general place name which embraced not only the island landing, but often Batsto, and particularly the village of Pleasant Mills, which was about a mile and a quarter northwest of the location mentioned by Heston.

Pleasant Mills today consists of a dozen or so houses, an historic mansion, a church erected in 1808, its adjoining graveyard, and an old paper mill which has been rebuilt into a picturesque theater. It is located about seven miles from Hammonton, between the Mullica River and the shore of Lake Nescochague, in Mullica Township, Atlantic County. For perspective on the history of Pleasant Mills, it may be best to go back to 1645 when Eric Mullica, a Swede, sailed up the river which was to bear his name. Charmed by what he saw, he established at or near Lower Bank the first white settlement in the area. In Mullica's time the Indians named Lenni Lenape roamed this region and are said to have had a "summer village" about eight miles farther up the river, at the

present location of Pleasant Mills. They called it Nescochague.[4]
Both the lake and a nearby stream now bear that name.

In 1707, when the first white settlers came to Nescochague, the
Lenni Lenape were friendly. The newcomers were families of
Scottish exiles. In 1685 they had fled the brutal war which Charles
II waged upon the Kirk of Scotland and at first had found refuge
among the Quakers in southern New Jersey. Striking out on their
own, they arrived at what was to become Pleasant Mills. Tradition
has it that the first building erected by them was a church, of
"unhewn logs, floor of clay and covered first with a thatch of
dried grass and afterwards with a roof of clapboards."[5] This
church is said to have been completed in a single day. It remained
in use about fifty years. Log cabin dwellings were then constructed,
and before long a village began to take shape. For a time these
hardy settlers—squatters though they were—lived by hunting,
fishing, and primitive agriculture. Game was abundant. There
also was trading with the Indians and later with nearby communities.

During the 1750's the neighborhood of "The Forks" enjoyed
considerable development as a trading center. Its location—on a
navigable river, yet inland and remote—made it a particularly
welcome haven for smugglers, who in those days did a thriving
business at quite a few points along His Majesty's coastline. More
sawmills began to spring up. One of the first had been built in
1739 by Samuel Cripps at a location not far behind the present
church.[6] Another mill, possibly legendary, was established by a
Jack Mullin near the present pond at Pleasant Mills. These mills
provided material for more new houses and stimulated a thriving
trade in lumber. Jersey pine and cedar were in great industrial
demand, and Jersey cedar, according to Tuckerton's historian, Leah
Blackman, was used for paneling and other interior woodwork in
"many of the finer homes of New York." One tradition has it that
the saws from Mullin's mill were taken to make swords for the
colonial cavalry in the Revolution. It is, at any rate, a romantic
idea.

Elijah Clark was the first "first citizen" of Pleasant Mills. The
picture of him which emerges from the past is that of a man of
stature, gentle in manner, forceful in purpose, blessed with educa-

tion and considerable culture, shrewd in business, and devout in his faith—an extraordinary character in many ways.[7] He was a native son. His birthplace was nearby, the oldest white settlement in Atlantic County, and still called, for his family, Clark's Landing.

Thomas Clark, Elijah's father and a colorful individual in his own right, had left Saybrook, Connecticut, to seek his fortune in Nova Caesarea, as New Jersey then was named. In the course of the search he reached the Mullica basin, and he liked it. He particularly liked a spot on the shore about seven miles below Pleasant Mills and roughly an equal distance from the point where the river flows into Great Bay. There, with his wife Hannah, Thomas Clark established both a home and a town. By 1718 Clark's Landing counted about forty dwellings, a trading house, and a log church. The population is not known, but it soon was increased by four—the sons of Thomas and Hannah Clark, whom they named Thomas, Jr., David, Samuel, and Elijah.

A delightful story concerning old Thomas Clark gives some idea of his quality. It appears that after the death of Hannah Clark the thoughts of the elder Thomas reverted to an old flame—perhaps his first sweetheart—who still lived in Saybrook. In the manner of Miles Standish he summoned his son, Thomas, Jr., then about 18, and charged him with a mission. He was to ride to Saybrook to "bring back Ruth that he might marry her, and if she would not come, to bring someone else who would consent to be his wife." Off went Thomas the younger on horseback for Saybrook, taking with him another horse, saddled, to bring back a bride whom the son might be choosing for his father! Just what Thomas, Jr., told his prospective stepmother is not known. Probably he did not mention that almost any girl in Saybrook would do. In any case, when Ruth answered "Yes," there was no need to shop about for a substitute.

It is recorded, however, that while in Saybrook young Thomas did do a bit of shopping about—on his own behalf! He met and quickly fell in love with the "beautiful and brilliant Sarah Parker." What Saybrook girls had in those days that others lacked, history does not reveal. After the long homeward trek with his prospective stepmother, and the wedding of the elder Thomas Clark, the boy hurried back to Saybrook and claimed his own bride. On their honeymoon, and on the way back to Clark's Landing, the bride and groom stopped off in New York, where young Thomas bought Sarah a set of "guinea gold beads and ear drops" for a bridal gift.[8]

Elijah, the youngest son of Thomas Clark, was not quite thirty when he decided to settle down at Pleasant Mills. From his brothers, Thomas and David, he purchased the fifty-acre tract on Lake Nescochague which later became his home and the seat of his plantation. This land was acquired by two deeds, one dated July 29, 1757, the other November 5, 1762. Two adjoining tracts —about twenty acres—were surveyed to Elijah Clark by the Council of West Jersey Proprietors.[9] This gave him a total of approximately 70 acres in the immediate vicinity.

In 1762 Clark built his homestead. For some years this has been called the "Kate Aylesford House," after the heroine of the novel, *Kate Aylesford*, by Charles J. Peterson, which was published in 1855. It is ironic that a once-popular piece of fiction should have so distorted actual history that not only has Elijah Clark been largely forgotten and his Revolutionary War prominence ignored, but none of the stories of the "Kate Aylesford House" even mention his name.

The Elijah Clark House—i.e., "Kate Aylesford House"—is a typical colonial mansion charmingly situated amid giant buttonwoods, with its broad lawn half-encircled by the quiet waters of Lake Nescochague. Some original clapboards have been replaced, one wing has been added, but its handsomely restored basic structure appears little changed since Clark's time. There is an ample center hall, with a staircase having the usual reverse turn at a landing. In the main wing there are four rooms on each floor, all equipped with fireplaces, which are served by the four double chimneys. On the lower floor several of the fireplaces have firebacks which are cast in a mold usually associated with Batsto Furnace and which almost certainly were made there. Two of the upper rooms have been paneled in the native cedar, and in these the fireplaces are framed in distinctive tiles, each of which is named and has a design of special significance, no two of them being alike. There are verandas both in front and at the back, the latter overlooking the lake. This venerable mansion has all the serenity of age and that patina of beauty which age bestows upon such of man's works as prove themselves worthy to survive. Elijah Clark's house stands today as it has through the years, stately evidence of those deep roots which bind one remote hamlet to the nation it helped build.

While Elijah Clark was establishing his homestead he also erected a new church to replace the old one, which had outlived its primitive usefulness. For many years thereafter the new church was known as "Clark's Little Log Meeting House." It was made of finished logs, with great red cedar beams, a clapboard floor, and a roof of hand-hewn shingles. That the meetinghouse was erected early in 1762 is indicated by the journal of the famous travelling preacher, John Brainerd, who for many years was a close friend of Elijah Clark, and was brother of David Brainerd, whose work among the Indians John carried forward. Brainerd records that on April 26, 1762, he "preached for the first time in the new meeting house."

This meetinghouse stood upon virtually the same site occupied by the present church, which bears the date "1808." Few records of its early days remain, and most of those which survive are in diaries of the various itinerant preachers who like Brainerd covered amazing distances on horseback to fulfill remote missions in that primitive territory. The Reverend Philip V. Fithian, of Cohansie, another frequent visitor, notes in his journal that he arrived there on Monday, February 6, 1775, and states:

> I rode to the Forks of Little Egg Harbour and put up according to direction at Elijah Clark's. Mr. Clark is a man of fortune and taste. He also appears to be a man of integrity and piety, an Israelite indeed. And, O Religion, thou has one warm and unfeigned advocate in good and useful Mrs. Clark. . . .

For Wednesday, February 8, Fithian added:

> I preached in Mr. Clark's little log meeting house. Present about forty. I understand the people in this wild and thinly settled country are extremely nice and difficult to be suited in preaching. One would think that scarcely any but a clamorous person who has assurance enough to make a rumpus and bluster would have admirers here. It is however, otherwise. They must have before they can be entertained good speaking, good sense, sound divinity, and neatness and cleanliness in the person and dress of the preacher.

Other accounts confirm that Elijah Clark's mansion was "in the wilderness a lodging place of wayfaring men" as Jeremiah put it. There hospitality was offered regularly to the visiting clergymen, a custom which endured well into the twentieth century. Clark and the early congregation were Presbyterians, but when the present edifice was built it became a Methodist church, although it is dedicated for use by "any Christian denomination."

Elijah Clark was born in 1730, and according to family records he was educated at Yale. He married Jane Lardner, also "of Gloucester County," who was about ten years his junior. The date of their wedding does not appear, but it probably was about the time that he decided to settle in Pleasant Mills. According to a granddaughter, Elijah Clark was a man of mind, taste and cultivation, a judgment confirmed by Fithian. He was an extensive reader, and his mansion housed a fine library. He owned many slaves, and he is said to have instructed and cared for them as if they were his children, which further attests to his having been a kindly and well-balanced, as well as a wealthy, man.

During the French and Indian Wars Elijah Clark was an officer in the colonial forces, and in due course he became a stalwart soldier of the Revolution. He served as a Representative from Gloucester County to the Provincial Congress of New Jersey in 1775. The following year—the momentous year 1776—he became a member of the convention which met in Burlington, Trenton, and New Brunswick from June to August. During the early years of the struggle for independence Clark was Lieutenant Colonel of the Third Battalion of Gloucester County militia, of which Pleasant Mills' other prominent Revolutionary figure, Richard Wescoat, was major. Late in 1777 Elijah Clark resigned his military post to become a Member of the Assembly, but he continued to be active and vigilant as the British made one menacing gesture after another in the direction of Little Egg Harbor and "The Forks."

Perhaps Elijah Clark had an added reason for his Revolutionary vigilance, for he has been credited, rightly or wrongly, with owning or having an interest in more than one of the privateers which sailed regularly from "The Forks" to prey on British coastal commerce. In any event, Clark became an increasingly wealthy man. When he purchased the Pleasant Mills tract he acquired the old sawmill. Soon afterward he established a gristmill, and later several other enterprises. They all appear to have flourished. As the years passed he purchased more and more parcels of land, many of them in the vicinity, but some as far away as Absecon and Mays Landing. When Clark sold these holdings they included 23 tracts—exclusive of the one in Absecon—and realized 18,000 pounds "lawful money of New Jersey," which was probably the equivalent of $100,000—wealth, in those times.

It is impossible today to ascertain the reasons why Clark, in 1779, decided to sell his properties. A farseeing man, he may have envisioned the postwar collapse of "The Forks" as a trading center. In any event, he has given us a good description of his own plantation in the following advertisement, which he placed in the *Pennsylvania Packet* for January 2, 1779:

TO BE SOLD

At The Forks of Little Egg Harbour
in Gloucester County, New Jersey

The premises whereon the subscriber now lives, with all the buildings and improvements thereon, to wit: A saw-mill and grist-mill, both remarkable for going fast and supplied with a never-failing stream of water, the mills within one mile and a quarter of a landing to which vessels of seventy or eighty tons burthen can come, skows carrying seven or eight thousand feet of boards go loaded from the mill; there is sufficient quantity of pine and cedar timber to supply the saw-mill for a great number of years, and also a great quantity of cedar timber fit for rails near the river side, which may be easily exported to those parts of the country where they will sell to great advantage; there is also on the premises, a dwelling house that will accommodate a large family, a barn, stables, and out-houses, also a number of houses for workmen and tradesmen, a smiths shop, wet and dry goods stores, and indeed every building necessary and convenient for carrying on business and trade extensively, for which the situation of the place is exceedingly well calculated, both by nature and improvement. Any person inclining to purchase may be more particularly informed by applying to the subscriber on the premises.   ELIJAH CLARK

Clark's mention of the landing serves to locate it in approximate relation to his plantation and "dwelling house." His description of the various other structures indicates that some of them were well outside the bounds of the little community proper. Just as today there is no trace of the buildings once located at the landing, it can only be conjectured that perhaps most of the commercial establishments mentioned were nearer to the landing than they were to the mansion house. This description, however, clearly places the mills close by the house, a fact confirmed by the old deeds and by the dependence of the mills upon waterpower from Lake Nescochague.

Even before the Revolution, Pleasant Mills and "The Forks" were of sufficient importance to justify a stagecoach stop. The first stage to Pleasant Mills—at least the first of which the writer can find any record—was established on March 24, 1773.[10] That day William McCarrell notified the public that he had "fitted a stage waggon to go from . . . Ann Risley's at Abseekam, on Monday morning, to go by Thomas Clark's mill and The Forks to Blue Anchor; from thence to Long-a-Coming and Haddonfield, to arrive at Samuel Cooper's Ferry [Camden] in the afternoon." The fare was one pence ha'penny per mile, and "for laying out cash for dry goods and other articles, one penny per shilling."

How long this stage service survived is not known. Five years later, however, another line was put in operation. The new operator was Samuel Marryotte, and he advertised as follows in the *Pennsylvania Packet* of September 3, 1778:

> To the PUBLIC: A Stage Waggon will set out on Monday morning from Peter Well's, at the Landing at Big Egg Harbour and to go to the Forks of Little Egg-Harbour, and from thence to Samuel Cooper's Ferry on Tuesday evening; On Thursday morning to set out from Samuel Cooper's Ferry, and to go to the Forks of Little Egg-Harbour, and from thence to Peter Well's at the Landing at Big Egg-Harbour. Those Ladies and Gentlemen who please to favour me with their commands, may depend on their being executed with fidelity and dispatch, by their much obliged humble servant,
>
> SAMUEL MARRYOTTE

Elijah Clark, as has been said, offered his plantation for sale in January of 1779. On April 2 of that year it was purchased by Richard Wescoat—the same Richard Wescoat who as a Major in the militia had been Clark's revolutionary comrade-in-arms. The deed to Wescoat, covering 23 tracts, specifically mentions "the Plantation whereon the said Elijah Clark now dwelleth . . . with a dwelling house, saw mill, grist mill and sundry other buildings thereon erected." After the sale Clark moved to the "Hinchman Farm," near Haddonfield, where he died in 1795. He is buried in the Old Presbyterian Cemetery in Woodbury, and his headstone reads:

> To the Memory of Elijah Clark
> who departed this life
> on the 9th of December 1795
> aged 65 years

> In memory of Jane Clark
> Relict of Elijah Clark
> who departed this life
> August 10, 1804
> in the 66th year of her age.

Richard Wescoat, second owner of the Pleasant Mills plantation, was born in 1733, the son of Daniel Wescoat and Deborah Smith Wescoat. The family was of English descent and had first settled in New England, later moving to New Jersey. The name itself is variously spelled, even in official sources, in some instances appearing as "Wescott," in others as "Westcott," whereas copies of his signature indicate that he himself preferred "Wescoat."

The young Richard seems to have been something of a dashing figure, a fit hero for historical novelists, and the pace he set in his early days he kept up far into his later life. He was a very early volunteer in the Revolution and was badly wounded in the Battle of Trenton.[11] After his recovery his wartime activities were confined largely to defense of the Little Egg Harbor area, as Major of the Gloucester County militia and later, after Elijah Clark's departure, Colonel of that organization.

Prior to his purchase of the Clark plantation, Wescoat lived near the landing at "The Forks," at a location still in doubt although old deeds place it about half a mile southeast of the Elijah Clark—or "Kate Aylesford"—house. Wescoat, as noted in an earlier chapter, had married a widow, Margaret Lee, by whom he had six children, and their descendants are numerous in Atlantic County today. Not far from Pleasant Mills itself is a little village called "Wescoatville."

Some historians say that Richard Wescoat kept a tavern and store at "The Forks." Since he was a wealthy man—one who could produce 18,000 pounds to purchase Pleasant Mills—it would appear obvious that he had some other sources of income. One of these was his interest in the nearby Batsto Iron Works. Another was the privateering trade which centered in that locality. Revolutionary archives tell of many privateers which were sold, with their cargoes, "at the House of Mr. Richard Wescoat" at "The Forks." Just as Tuckerton and Little Egg Harbor Bay were what the British called a "nest of rebel pirates," so "The Forks" had become an even more secure haven for privateers of lighter draft. Here ships unloaded barrels of sugar and bags of coffee, boxes of tea, puncheons of rum—to be hauled overland to Philadelphia, usually through the old Pine Barrens trails to Atsion and thence by way of Long-a-Coming and Haddonfield. While Philadelphia was occupied by the British the wagons went to Lumberton, in some cases all the way to Burlington, where the shipments were ferried over the Delaware beyond reach of the redcoats. Along the pineland trails there were said to be many hiding places, with caches for goods in every bog, while loads of salt hay might conceal sugar, molasses, and other contraband, for even in the pines there were Tory spies with prying eyes. It was Richard Wescoat, the records indicate, who organized that traffic, so valuable to the colonies.

Among the many vessels brought up the Little Egg Harbor River to be "sold at public vendue" at "The Forks" were two of particular interest: the *Black Snake*, a brig mounting eight-pounders, and another vessel, *The Morning Star*. Behind their sale lay one of the most exciting stories of the Revolution as it was fought off the Atlantic coast.[12] The *Black Snake* was an American privateer which had been captured by the British fleet. Captain William Marriner, of New Brunswick, proposed to recapture it. The measure of his daring is that he set out with all of nine men —and a whaleboat! Captain Marriner not only retook the *Black*

Snake, he then put to sea in her. Coming upon the British war-
ship *The Morning Star,* he captured that as well. This feat was
accomplished in spite of the fact that *The Morning Star* was
armed with six swivel guns and carried 33 men. In the battle three
British seamen were killed and five wounded. Then the brilliant
Captain Marriner—so well named—brought both his prizes to "The
Forks" for sale.[13]

Among other vessels similarly sold was the sloop *Dispatch* (or
Speedwell), captured by Captain William Treen in the schooner
Rattlesnake. Also, in May, 1780, "at the House of Mr. Richard
Wescoat," Zachariah Rossell, then Marshal, sold "The sloop
SWALLOW, burden about 70 tons, with four three-pounders and
four swivels; together with all her tackle, apparel and furniture;
also pork, beef, bread, powder, ball &c. captured by Capt. Nathan
Brown and others."

Richard Wescoat was involved in a still-unsolved mystery concerning the famous Battle of Chestnut Neck, near the mouth of
the Mullica. Early in 1777 Elijah Clark and Colonel John Cox,
then owner of Batsto Iron Works, petitioned the Council of
Safety for help in fortifying Chestnut Neck, after warning had
come of a projected British raid. Their petition relates that

> We have presumed to take from Capt. Shaler eight or ten pieces
> of cannon . . . with orders to throw up a battery to defend the
> Inlet. There are now . . . a guard of about 20 men, and Col.
> Clark will immediately order down as many more to assist in
> doing the necessary work. Powder and provisions will be im-
> mediately wanted. Shot can be procured here.[14]

Later that year the Assembly was notified that Clark and
Richard Wescoat had erected, "at their own expense," a small
fort at Chestnut Neck, and reimbursement was requested. It was
granted in September at an Assembly session in Haddonfield. The
mystery, however, surrounds the fate of those cannon which Clark
and Wescoat had provided. For when the British made the attack
which destroyed Chestnut Neck the following year, its quick fall
was attributed to the defenders' lack of artillery, and British
accounts of the battle contain mention of "Scotch" Ferguson's
surprise that the cannon emplacements were empty.

In those years Richard Wescoat also was in charge of govern-

ment stores at "The Forks," an important post considering the
volume of trade there. On December 15, 1779, he wrote, in the
line of duty, the following interesting letter [15] to William C.
Houston, Member of Congress, Philadelphia:

> Sir—I have the satisfaction to inform you that I have been down
> to Absecon Beech and have gott all the wine belonging to the
> United States landed on the said Beech Excepting four or five
> Casks which were bilged and almost out. The Severity of the
> Weather and exceedingly high winds which prevailed have ren-
> dered it out of my power to Bring it to the main land which is
> seven or eight miles from the Beech. The Scollops which I ex-
> pected to take it in have not yet arrived. I shall therefore have it
> properly stored and taken care of till opportunity presents of con-
> veying it to the City of Philadelphia.
>   RICHARD WESCOAT

Now comes the story of *Kate Aylesford*, the melodramatic novel
by Charles J. Peterson, which has Pleasant Mills and the old Clark
mansion as its locale and the Revolutionary events of the Mullica
basin as its background. This is a work which must be savored
slowly to be appreciated.

Kate is a heroine in the most extravagant Victorian tradition.
She is the quintessence of all feminine virtues. Her beauty beggars
description. Her lovely voice "soars" above all others in utter
purity when hymns are sung in church. She is an only child, an
orphan, and one of the richest heiresses in all the American colonies, a combination which evokes sympathy on the one hand and
excites envy on the other. She is shipwrecked, faced with death,
rescued by her hero, Major Gordon, trapped by a forest fire, kid-
naped, tracked by a bloodhound, rescued again. And when her
romance finally conquers all and is crowned with marriage, she is
given in wedlock by no less a personage than General Washington
himself.

Curiosity has long existed as to the identity of the prototypes—
if any—of Kate herself, her heroic lover, Major Gordon, and other
characters in the novel. There has been much guessing on these
matters over the past century, quite a bit of it in print, with a tend-
ency in some quarters to accept the novel itself as a paraphrase
of history. At the other extreme certain skeptics have questioned

whether Kate—or any such paragon of a woman—ever existed at all!

There was indeed a real-life Kate. She was not, however, "Honoria Reid," who some have suggested was a daughter of Charles Read, the famous ironmaster of Batsto. Charles Read had no daughter. The real "Kate" was Margaret, youngest daughter of Colonel Richard Wescoat, known to have been, in all truth, lovely, vivacious, charming, and intelligent. Margaret Wescoat lived with her parents at "The Forks." According to the family records she was sent to a private school for "young ladies" in Burlington, which was conducted by an Englishwoman named Davenport. At Burlington Margaret Wescoat at least saw General Washington several times. On one occasion her entire school was taken to hear him address the Army in the Public Square. History is vague on Margaret Wescoat's first marriage, to a youth named Leonard. She must have been very young. Apparently it was a war marriage, and it is a reasonable presumption that her husband died in service. The name of an Azariah Leonard appears in the records of the old Gloucester County militia, but whether he is the Leonard in question remains unconfirmed. In any event, Margaret was a widow—at 18—when the romance which resembles that of Kate Aylesford took place.

Margaret Wescoat had never heard of Nathan Pennington until the day he rode up the Mullica River—then the Little Egg Harbor River—reportedly on the prize ship of a privateer. At that time Nathan Pennington was in charge of captured property at Chestnut Neck. Previously, however, he had had a colorful career. At 19 he had volunteered in the Army. After he had seen some hard fighting he had been captured by the British, and, with others of His Majesty's prisoners, taken to Quebec for "safe keeping." Pennington, however, was not an easy man to hold. Soon he escaped, eluded pursuit, and in due course turned up in New Jersey, where he was given the assignment at Chestnut Neck.[16]

Pennington's trip up the Mullica that special day started as a business matter, for he was to consult with Colonel Wescoat. That he also "consulted" with Margaret Wescoat is history. A wartime courtship followed, they were married, and after the war Nathan became a shipbuilder and took up residence in Mays Landing at what long was called "Pennington's Point." Margaret and Nathan Pennington had four sons and five daughters. It is interesting to note that Nathan's brother, William S. Pennington, was Governor of New Jersey from 1813 to 1815.

Because *Kate Aylesford* has both obscured and confused the real history of Pleasant Mills, it may be useful to indicate how it mixes the fiction and the facts. Charles J. Peterson, whose grandfather was one of the original Trustees of the Pleasant Mills Church, wrote his novel in 1855. In the manner of novelists, Peterson drew some of his story from history, but he scrambled that history as one might scramble eggs. Thus one character in the novel may possess the attributes of two actual people—or more —who lived in or about Pleasant Mills. Members of different families are combined for fictional purposes into one family, and their children are casually transferred to another.

To begin with, the name "Sweetwater" dates from the novel. It never was applied to Pleasant Mills in any deed, agreement, or other contemporary record known to the writer, nor was the lake originally called "Nescochague," from which the term supposedly derives. It was "Clark's Mill Pond" for many long years, and the Nescochague Creek—which appears in old deeds as "Neskeetchey" and "Echeocheague"—does not empty into the present Lake Nescochague, but joins the Atsion River back of the old church and graveyard. "Kate Aylesford" herself, as has been shown, almost certainly was Colonel Wescoat's daughter, and the novel describes her excellent education, the wealth of her family, and her residence at Pleasant Mills and the old homestead. Margaret Wescoat, however, was not an orphan, and the shipwreck and kidnaping incidents, had they occurred, would surely have been mentioned in the family records! "Major Gordon's" fictional history also bears some relation to the career of Nathan Pennington: both volunteered in the colonial cause; both were captured by the British; and both were at Chestnut Neck.

"Uncle Lawrence Herman," a farmer and philosopher in the novel, is quite clearly patterned after Simon Lucas, farmer, lay preacher, and captain in the Gloucester County militia. Like his fictional counterpart, Simon Lucas is buried in the old Pleasant Mills cemetery. However, author Peterson allots to "Uncle Lawrence Herman" the war wounds which were suffered in the Battle of Trenton, not by Simon Lucas, but by Richard Wescoat, father of the bride. The villain of the novel, "Ned Arrison," is a fairly close portrayal of that notorious "Refugee" bandit, Joe Mulliner, whose career is described below. Some of Mulliner's actual exploits are worked into the book, but while he had a dog, it was a small one, not a bloodhound such as the panting beast which traps Kate Aylesford in a cedar jungle near "The Forks."

A final mingling of fact and fiction concerns the children of "Major Gordon" and "Kate Aylesford." In the novel they all become important people or marry important people—one a Senator, another an Ambassador, and so on. In real life such marriages were made by the daughters of Colonel John Cox, wartime owner of the nearby Batsto Iron Works.

As fantastic as any fiction were the Revolutionary activities of the so-called "Refugees." Supposedly loyal to the Crown, they actually were gangs of ruthless thugs, who ravaged the countryside, held up stagecoaches, burned and pillaged farmhouses, and terrorized helpless housewives while their menfolk were off fighting the British.

Some of the worst of these gangs operated along the Little Egg Harbor River. One was led by the famous Joseph Mulliner, who sometimes has been pictured as a happy rogue with the warm heart of a Robin Hood. Actually, he was nothing of the sort. This tall, handsome, swaggering Englishman was a treacherous, crafty scoundrel, albeit a colorful character. It is said that he always wore a uniform, carried a sword, and packed a brace of pistols. And walking arensal though he was, he prudently surrounded himself with bodyguards. Indeed, Mulliner's gang is said to have numbered as many as a hundred men, most of them mere wartime riffraff and a foul crew by any standard.

The "Refugees" apparently parceled out territory among themselves much as latter-day gangsters have done. Mulliner's stamping ground was "The Forks," and his headquarters is said to have been an overgrown and almost inaccessible island in the river at a spot called Cold Spring Swamp. Around this hideaway remarkable tales have been woven. One relates that he built boats camouflaged with small cedars, so that the craft could be maneuvered along the Mullica with minimum d nger of detection. Another legend concerns Mulliner's dog, which he had trained as a courier. While Mulliner's headquarters was on the island, his wife—a fitting help-meet—did much of the spying for the gang from a cabin on the mainland. To communicate with her, especially when Patriot forces were in the vicinity, the bandit fashioned a dog collar with a secret compartment. Whenever he chose to send a message he would conceal it in the collar, then start the dog swimming across the river. After all, who would suspect a dog?

In his *Pleasant Mills*, the late Charles F. Green has given this account of a raid by Mulliner's gang on the farm of a widow named Bates.

> She was a large masculine appearing woman of fearless disposi-tion and an ardent patriot. Of her eight sons, four were serving in the American Army. The others being too young for military duty assisted her in tilling their farm. Among her household pos-sessions were some pieces of rare old furniture and a service of silver plate which was highly prized as a family heirloom.

Returning from the Little Log Meeting House one Sunday, Mrs. Bates found Mulliner's gang in possession of her house. For some reason Mulliner was not with them. They had ransacked the premises and were carrying off, among other things, the precious service of silver plate. This was too much for the Widow Bates. She gave the gang "a terrific tongue lashing."

"Silence, woman!" cried the leader of the gang, "or we'll lay your damned house in ashes."

"That would be worthy of cowardly curs like you," snapped the widow, "But you'll never stop my mouth while there's breath in my body."

Thereupon the Refugees took firebrands from the hearth and set the house ablaze. When Mrs. Bates attempted to douse the flames with buckets of water, while her children pelted the gang with rocks, the brigands bound the youngsters with ropes and lashed the widow herself to a nearby tree. There she watched help-lessly while her home burned to the ground. As the story goes, friends came to her assistance in great numbers, and several weeks later she received an anonymous donation of $300. Some said it was from Mulliner, to atone for the outrage committed by his gang. Others—more realistic, it would seem—called this "a likely story."

Legends inevitably cluster about a character such as Joseph Mulliner, and many of them are probably as short on fact as they are long on fancy. He does seem however, to have been a gay blade in his spare time, and many of the legends concern his reputation for crashing dances and parties, sometimes to stage a holdup and on other occasions merely to make merry. Indeed Mulliner's dancing hobby seems to have proved his undoing. When he finally was captured, reportedly by Captain Baylin's rangers, it was at a dance in what is now Nesco.[17] Trapped amid the festivities, Mulliner was seized, taken to Burlington, and imprisoned. Six weeks later he was tried, found guilty, and sentenced to be hanged. It is said that Mulliner was placed in a wagon along with his own coffin and taken to a spot called Gallows Hill. Crowds followed the grim equipage. The *New Jersey Gazette* of August 8, 1781, reported:

> At a special court lately held in Burlington, a certain Joseph Mulliner of Egg-Harbour, was convicted of high treason and is sentenced to be hanged this day. This fellow had become the terror of that part of the country. He had made a practice of burning houses, robbing and plundering ALL who fell in his way, so that when he came to trial it appeared that the whole country, both Whigs and Tories, were his enemies.

Mulliner's body was sent to his wife, at Pleasant Mills, where he was buried. About 1850 his bones were dug up by a party of drunken woodsmen and taken to Batsto, but Jesse Richards had them returned to their original resting place.

Craftier and more fortunate than Mulliner was another notorious Refugee leader, William Giberson. Captured while hiding in the house of some girl friends at Little Egg Harbor (Tuckerton), he was taken off to Patriot headquarters. On the way he tricked his captor, Lieutenant Benjamin Bates, by diverting his attention, and then vanished into the night. Bates was ordered to recapture Giberson at all costs. He therefore returned the next night to the same house. One of the girls came to the door. As Bates was about to enter he heard the click of a gun. Turning, he saw Giberson aiming at him from behind a tree. Bates fell to the ground, and the shot passed through his hat. His swift counterfire, however, caught the fleeing Giberson in the leg; and although he managed to reach a nearby swamp, the "Refugee" was captured, taken to Burlington, and put in the same prison which had housed Joseph Mulliner.

Stone walls, however, did not make a prison for Giberson. Resourceful as well as ruthless, he asked permission to have his sister visit him in his cell. The permission was given. No one realized that the sister "bore a striking resemblance both in face and form" to her brother. During her visit to the prison the two exchanged clothes, and not only did Giberson leave the prison in woman's attire, but his jailer was so completely taken in that he helped the desperado into the wagon which waited outside. Giberson soon reached New York, joined the British Army there, and was last heard of in Nova Scotia.[18]

With victory in the Revolution not far off, and dwindling of war boom prosperity at "The Forks" in sight, Richard Wescoat decided to sell the Pleasant Mills Plantation. It was a shrewd decision, for local land values were then at their peak. The purchaser was Edward Black, of Mount Holly, and the deed, of March 18, 1782, specifically recites that Parcel Number One is "the Tract or Plantation whereon Elijah Clark formerly dwelt."

Richard Wescoat moved to Mays Landing, where he had acquired some property. Characteristically, when Wescoat moved in he moved "all the way," and before long he was close to being political boss of Mays Landing and certainly was one of its first citizens. For some years he ran a gristmill there on Babcock's Creek (the old Iliff mill). He also kept a hotel and a store, and the latter seems to have been an important trading center for some years, especially for the Indians. Wescoat's rapid rise in the town's affairs is shown by the fact that in 1795 he was Town Clerk, Surveyor of Highways—along with David Sayrs—and also one of the three "Pound Keepers." The beautiful oak trees which now grace the Main Street in Mays Landing are said to have been planted under Wescoat's direction.

Edward Black kept the Pleasant Mills plantation less than five years. Whether he ever lived there does not appear. The chances are that he did not, as he was a man of some prominence in

Quaker circles in Mount Holly, and most likely he operated the mills through a local manager. On January 22, 1787, Black sold Pleasant Mills to Joseph Ball, at that time one of the partners in the nearby Batsto Iron Works.[19] Through this purchase Pleasant Mills and Batsto were brought under common control for the first time, a situation which in one degree or another was to continue for many years.

In purchasing Pleasant Mills, Joseph Ball was making another of the many investments which were to establish him among the wealthier men of his day. Later he was one of the four partners who operated the Weymouth Iron Works. Ball also owned large tracts of land elsewhere in New Jersey, as well as in Pennsylvania, Virginia, and the District of Columbia. He was one of the original directors of the Insurance Company of North America, as has been noted, and a director or officer of several other leading Philadelphia financial institutions until his death, in 1821.[20]

One significant aspect of the Ball ownership of Pleasant Mills lay in the setting aside of the meetinghouse and cemetery as church property. Previously these had been part and parcel of the plantation itself, but when Ball sold Pleasant Mills to Samuel Richards and Clayton Earl, on August 24, 1796, the two acres occupied by the church and burial ground were excepted from the conveyance, title remaining in Joseph Ball.[21] In 1808 when the "new" and present church was built, Ball deeded the ground, on October 10, to its trustees: William Richards, Simon Lucas, George Peterson, Laurence Peterson, Gibson Ashcroft, John Morgan, and Jesse Richards. The following extract from the deed is interesting and does not appear to have been printed before:

The intent and purpose [is] that they hold the House now thereon erected and such other house as may hereafter be erected thereon as a House or place of public worship for the use and enjoyment of the ministers and preachers of any Christian denomination to preach and expound God's holy word, in particular for the use and enjoyment of the Traveling and local ministers of the Methodist Episcopal Church . . . whenever and as often as they may appoint to preach and expound God's holy word or to hold occasional meetings of the Society therein without any let or molestation and in further trust and confidence that the said trustees or a majority of them whenever they shall deem it expedient may

erect and build a School House on the hereby granted premises
for the use, benefit and enjoyment of the inhabitants of that and
the adjacent neighborhood.

It is interesting that Joseph Ball envisioned a schoolhouse at
Pleasant Mills and made this provision for its possible erection. A
school was in operation some sixty years later, but no advantage
was taken of this proviso, and the classes were held elsewhere.
More than likely Ball's proposal had been forgotten.

Clayton Earl, who bought Pleasant Mills jointly with Samuel
Richards, was another speculator. He was keenly interested in iron,
once having held an interest in the Hampton Forge northeast of
Atsion. He also is listed as one of the builders of the Hanover
Furnace. Earl retained his ownership of Pleasant Mills less than
two years, conveying it to William Richards on February 28,
1798.[22] Samuel Richards' part ownership was purely nominal. On
October 26, 1796, two months after he and Earl took title, he
signed his share over to his father. That fact was not officially recorded, however, until thirty years later, two years after William
Richards' death. Thus the "Lord of Batsto" was the real owner
of Pleasant Mills for a quarter of a century, even though he did
not live there and after 1809 made his home in Mount Holly.

Simon Lucas, who has been mentioned previously in connection
with *Kate Aylesford*, was another of the remarkable characters of
Pleasant Mills in those early days. A farmer, captain in the militia
and rustic philosopher, Lucas is said to have "got religion" from
John Brainerd. In any case, he became a lay preacher and seems to
have conducted most of the services at Pleasant Mills until his
death in 1838.

Many tales are told about Simon Lucas.[23] One of the best concerns the day he stepped into the pulpit to announce a hymn and
saw before him, in the front pew, a strange young woman who was
very flashily dressed. Among other things, she wore a glittering
brooch.

"Young woman," said Simon Lucas, "Do you know, that shiny
thing on your dress reminds me of the devil's eye!"

Blushing and angry, the girl hastened from the church. The
service went on quietly. Simon Lucas had seen his duty—and
done it!

Another story suggests a more flexible attitude toward his flock. It seems that one Sunday a large run of herring appeared in Atsion River, and the male population turned out en masse. Charles F. Green writes:

> While the sport was at its height and both banks of the stream were covered with flopping fish, Jesse Richards and one of his daughters passed by. The young lady was deeply shocked.
>
> "Papa," said she, "why don't you stop those men from fishing on Sunday?"
>
> "Don't know that I have any right to," replied the old gentleman, "but I'll ask Simon Lucas what he thinks of it."
>
> On their way home from church they found the fishermen still busy and Miss Richards said: "Papa, did you ask Mr. Lucas about Sunday fishing?"
>
> "Yes," said her father. "I mentioned it."
>
> "And what did he say?"
>
> "Well, he said the time to catch herring is when the herring are here to catch."

This same story has been told in various other connections; a variation of it comes from Bermuda, where it is linked with one of the early rugged pastors of St. George's. It is at any rate a good story, and it reflects the mores of such a tight little community as Pleasant Mills was at that period.

There is no doubt at all that Simon Lucas was highly regarded by all who knew him. That esteem was tangibly expressed in property deeds which noted "exceptions" granting Simon Lucas the right to use a "hay landing" even though it was on someone else's land along the Mullica River. Both Simon Lucas and his wife Hannah enjoyed long lives—he died at 87 and she at 83. The great stone on their grave in the Pleasant Mills cemetery notes that they were members of the church for more than sixty years and "were distinguished as most exemplary and zealous followers of Christ."

In the same little graveyard are buried many members of the Richards family, including Jesse Richards, master of Batsto, and Mary Patrick, the first wife of ironmaster William Richards. Near that famous family lies another—the Wescoats—including Daniel and his wife, Deborah, both of whom died in 1791, the former at

83 and the latter at 82. Another family still remembered is that of the Petersons, eight of whom are buried nearby; and the Sooy family is represented by the famous Nicholas, who was host and founder of the famous tavern at Washington, five miles away. One of the most interesting headstones reads: "Here Lies the Body of Capn. John Keeny who died January the 6th 1760 in the 34th year of his life." Keeny is said to have been a veteran of the French and Indian War and a comrade-in-arms of Elijah Clark, whose guest he was at the time of his death.

Inside the church are two memorial tablets. One was placed there on November 14, 1914, by the Kate Aylesford Chapter, D.A.R., of Hammonton, in memory of the Revolutionary soldiers and sailors buried in the old graveyard. The other tablet was dedicated on November 10, 1935, to the memory of Charles F. Green, who long lived in Pleasant Mills and wrote the first historical sketch about it, already mentioned above. In 1908 Green wrote a pamphlet (which does not bear his name) on the history of the old church, and in it notes, "The original pulpit was taken down about forty years ago and the old fashioned seats replaced by others of a more modern type. The removal of the pulpit was regarded by some of the older residents as an act of sacrilege." Green also lists many of the ministers who preached at Pleasant Mills church, and these included, to name but a few, John Core, J.J. Hanley, Samuel Downs, Nicholas Vansant, Peter Burd, C.A. Malmsbury, William Van Horn, W.L. Peterson, Nathan Wickward, W. Vanderhurschen, and J.R. Mason.

It was the Reverend C.A. Malmsbury who recorded, in his *Life of Charles Pitman*, D.D., the Presiding Elder of the West Jersey Methodist District, who often was at Pleasant Mills, a story of the throngs which would sometimes gather for the services there. He notes that

> Sometimes so great would be the crowd that Mr. Jesse Richards would send over his six-mule team open wagon, cover the bottom with planks and place it in front of the church to improvise a pulpit. The women and children would be seated in the Church, the doors and windows open, with the men standing outside. Mr. Pitman, standing upon the wagon, would address from two to three thousand people for an hour or more. . . .

Along the road which curves toward Weekstown the remains of yet another cemetery may be found. Lying about a quarter of a mile from the Elijah Clark House, this cemetery is easily missed, since it is a bit off the roadway and is all but hidden by trees and underbrush. It is doubly interesting, however, since it marks the site of a largely forgotten Roman Catholic church, the second oldest Catholic edifice to be built in the Diocese of Trenton. Completed in 1827, it was dedicated three years later and named "St. Mary's of the Assumption." Services were held there for thirty years.

There had been no Catholic church nearer than Philadelphia. Half a century before, occasional Catholic services had been conducted at "Shane's Castle," near Waterford, about ten miles away. There, as in Pleasant Mills, many of the parishioners were members of Irish families, whose breadwinners were employed in the various bog-iron works, sawmills, and other industries in the pines. "Shane's Castle" was actually a log cabin. Its site has never been definitely located. It was built by three brothers named Woos, who came from Germany, and it is said that for quite a few years services were conducted there more or less regularly by visiting priests. It has been thought by many that those were the earliest established Catholic rites in West New Jersey.

St. Mary's of the Assumption was a frame structure, built in large part by the parishioners themselves under guidance of their energetic pastor, Father Edward R. Mayne, who came down monthly from Philadelphia.[24] The site supposedly was donated to the church in 1826 by Jesse Richards, an Episcopalian, who employed most of the members of the congregation. It seems difficult to believe that the church would have been built without formal transfer of title to the land. Yet no record of a deed to St. Mary's has come to light. It is true that back in those days, and even in later years, people often were careless about having deeds recorded. In some instances transfers of considerable business property were not recorded, and in other cases the recording was not made until years after the original transfer of ownership. Correspondence of Alexander J. McKeone, long owner of the Elijah Clark House, with the late Bishop James H. McFaul in 1899 indicates that the Reverend Hugh Lane, a former pastor of St. Mary's, was of the opinion that a caretaker named Jerry Fitzgerald had kept the deed "because the church owed him some money." In the Clevenger Papers, held by the Atlantic County Historical Society, it is stated that a caretaker's house was once erected near the church and occupied by Fitzgerald. That some difficulty did exist with regard to him is suggested by the fact that when the priest visited Pleasant Mills he lodged with Jesse Richards, and that Richards' daughter, though of a different faith, took care of the altar. In any case, the matter of the title remains a mystery.

Regular care is now promised the neglected cemetery, and some kind of historic marker belongs there, for among the pastors of St. Mary's of the Assumption were such well-remembered priests as Father Mayne, who died at 34; the Reverend Edward McCarthy; the Reverends Cumnisky, Finnegan, Bayer, and Lane. The grave-yard itself gives the names of some of the parishioners, recorded here before vandals cause further damage to the old markers:

Joseph and Helena Albor; Aaron, Ann, and James Dellet; Lydia, Mary, John, Jr., James, James, Jr., and Daniel Delett; Daniel and Eliza Elberson; Franz and Mary Froehlinger; Daniel Kane; Martha Kirby; Cornelius and Mary Kelly; Ann and Margaret Langtry; Anthony and Agnes Long; Mary, Mary A., Ann, James, James, Jr., and John McCambridge; John and Francis McIntyre; James and Mary Milley; John McSwiggan; John Smith; Mary R. Wills; and Regina Woolman.

The oldest grave is that of Mary McCambridge, who died November 21, 1835, at the age of 69; the latest is that of George McCambridge, who was buried in 1906 at the age of 18. The monument to the Langtry sisters, Ann, 42, and Margaret, 50, was "erected by their brother Michael Langtry." The headstone of the Froehlingers is in German. They were the parents of Joseph Fralinger, who became prominent in Atlantic City as the "father of salt water taffy."

Failing fortunes in the nearby industrial enterprises had a swift effect upon St. Mary's. With the Batsto Iron Works closed forever and the Pleasant Mills cotton factory burned down, the last service in the church was held in December of 1860, with only eleven persons present. Most of the congregation had moved away, the men in search of employment, while a number had died. In

1865, when Father Byrne of Gloucester visited Pleasant Mills, the few Catholics still residing there chose not to assemble in the old church; instead, services were held in a private house.[25]

In 1866 St. Mary's was made a mission of St. Nicholas parish in Egg Harbor City. Later it was transferred to St. Joseph's, Hammonton. The church itself was burned to the ground in a devastating forest fire on April 27, 1900.

The Pleasant Mills plantation passed into the capable hands of William Richards in 1798 and remained wholly or partly in the Richards family for more than half a century. At the time that Richards acquired Earl's interest, he was busily purchasing additional tracts of land in the general area of "The Forks." The Pleasant Mills transaction carried with it not only a gristmill, a couple of sawmills, and several lesser enterprises, but also valuable ore and water rights. The Richards era in the Pine Barrens was just beginning, and the heyday of the bog-iron industry in general was not to arrive until the War of 1812.

William Richards and his sons were able and farsighted men. They recognized that, for all their wealth, most of their eggs were in one economic basket—the iron industry. So, seeking what today would be called "diversification," they moved to establish other enterprises in their feudal domain. For Pleasant Mills they chose to sponsor erection of a cotton factory, which was leased in a ten-year agreement, dated February 25, 1822, to Benjamin Richards—another of William's sons and later Mayor of Philadelphia— and to four partners.[26] One of these partners was William Lippincott, a brother-in-law of Jesse Richards. The others were Edward Yorke, William Yorke, and Benjamin Soy. The terms of the lease are interesting. The rent for the cotton factory was to be $250 a year. Included were "all that mill seat commonly called Pleasant Mills together with the mill stream and pond, the water rights to same, and two hundred acres of land adjoining." Also included in the lease agreement were two options. One of these provided that the five partners could purchase the mill outright for $4,000 at any time within five years. The other option covered the old Batsto Forge Tract, with a like sum to be paid within five years if they chose to buy. It so happened that William Richards' will did not empower his executors to fulfill this agreement, so that after his death had taken place in 1823 it became necessary to obtain a special act of the New Jersey Legislature to permit the terms to be carried out. This act was passed and became law December 1, 1824.[27]

William Lippincott seems to have been the most active partner in the first days of the cotton mill. Benjamin Richards was busy in Philadelphia, and Soy seems to have been little more than an investor. Edward and William Yorke sold out their interest to "Lippincott and Richards" on January 1, 1826. Soy sold out a year later. Still more changes were to come. Not long afterward William Lippincott died, and in 1827 his executors conveyed his "two-fifths part in the leasehold . . . buildings, factory and improvements . . . called Pleasant Mills" to Joshua Lippincott.[28]

Meanwhile, after William Richards' death his properties had been put up for auction, in 1824, at the Merchants Coffee House in Philadelphia,[29] and, as has been noted, most of the properties were bought by the family. Title to Pleasant Mills, Batsto, and 50,000 acres thereabouts went to Thomas S. Richards,[30] a son of Samuel and a grandson of the ironmaster. He was to retain a half interest in Batsto for many years. As for Pleasant Mills, on June 15, 1827, the old lease was terminated and the whole of the mill property deeded to "Joshua Lippincott, Merchant, and Augustus H. Richards, Attorney-at-Law." The grantors in this deed included Thomas S. Richards, Jesse Richards, and the "Executors of William Richards."

This transaction was made under the legislative authority already mentioned. It fulfilled the lease agreement, and Lippincott and Richards obtained the Pleasant Mills tract and the Batsto Forge Tract at the stipulated price of $4,000 each. Another significant aspect of the deed is that it specifically grants "the right and privilege of cutting and making a canal or race from and to the Batsto Forge Stream and Pleasant Mills Stream and thereby divert, lead and convey any part of the waters of either of the said streams to the other of them for the use of any mill or Mill Works. . . ."

Such a canal was dug and is still in existence. The "Forge Stream" mentioned appears on maps as Nescochague Creek, and it flows not into Lake Nescochague, as commonly supposed, but normally into the Atsayunk or Atsion River near the point where it used to be dammed to make the big Batsto Forge Pond. The purpose of this canal was to divert water from the Forge Pond to Lake Nescochague so as to boost the waterpower and maintain a more constant flow. The latter objective, however, was not always achieved, for in later years the Pleasant Mills factory installed a steam-powered standby plant for use when the waterpower was insufficient. The canal itself ran dry about 1940, when the Forge Pond dam collapsed and was not rebuilt.

Only five days after the transfer of the Pleasant Mills tract there was another real estate shuffle in the Richards family. Augustus sold his half interest in Pleasant Mills to "Benjamin W. Richards, Merchant." A significant exception in this deed reserves to

> Thomas S. Richards, his heirs and assigns the use of the Ores of Iron which might be found in and upon the premises at any time forever together with the right at all reasonable times of ingress, egress and regress for the purpose of collecting and carrying away the same. Provided, however, that the right should be exercised as not to expose the owners of the hereby granted premises their heirs or assigns to any serious loss or injury.

The purpose here was to assure the Batsto Iron Works the right to any bog ore which might be found on the Pleasant Mills tract, and while this exception carries down through the line of title to the present day, there is no evidence that the rights thus granted were ever exercised.

At the time Augustus Richards sold his share of Pleasant Mills to his brother Benjamin—on June 20, 1827—the tract comprised 380.85 acres "including the Pleasant Mills and Batsto Forge Ponds respectively." The portion touching the latter pond was called the "Batsto Forge Tract" because the first Batsto Forge had been located there (about half a mile from the furnace site near the Batsto mansion). No records of the cotton factory itself have come to light. It is known to have been a four-story stone structure. There were 3,000 spindles and a "large number" of employees. Given such a valuable property and presumably the backing of the Richards family wealth, Pleasant Mills might have been expected to prosper. That it was not doing so at this period of operation is suggested by the fact that on May 8, 1826, it had been mortgaged for $10,000.[31] The lender was John Moss, and the borrowers were William Lippincott, Benjamin Richards, and Benjamin Soy. This mortgage was to complicate the affairs of the cotton mill some years later. In 1832 it was assigned by Moss to John K. Kane and William and George Richards as "Trustees for Mary Richards," the young daughter of the ironmaster's declining years. When economic difficulties set in again, in the 1850's, foreclosure proceedings followed, and the net result appears to have been to eliminate the Lippincott interest from the ownership of Pleasant Mills. After the foreclosure was completed, title was conveyed by Richard W. Howell, Master, to Augustus Richards, and Pleasant Mills, after half a century, was once more wholly under the control of the Richards family.

In 1854, following the death of Benjamin, another member of that extensive family—Lewis Henry Richards—took over the factory and plantation. This was accomplished by two deeds, both dated November 13, 1854, one from the heirs of Benjamin Richards, the other a conveyance from Augustus, who thus passed out of the picture. Lewis Henry Richards' tenure, however, was very brief; after only one year it ended in disaster.

Fire was the nemesis of Pleasant Mills then, as it has been so often since. In July, 1834, there had been a serious blaze at the mill. The *Camden Mail* of July 25 reported that the plant had had a "narrow escape from conflagration." The account continues:

> The electric fluid communicated to the Loom House, which together with the Dye House were entirely consumed; other parts of the works were injured and the whole perhaps would have been destroyed but for the exertion of the neighbors a large collection of which was attracted thither to witness the performance of a traveling Circus company.

On November 27, 1855, fire struck again. This time the whole Pleasant Mills factory was destroyed. There must have been dismay in the hearts of the employees when it was decided not to rebuild. This decision probably was influenced by the closing down of the Batsto Iron Works only a short time before, an event which already had brought considerable economic distress to the locality.

Despite the economic ups-and-downs of the cotton mill, the community had expanded as more housing had been required for its employees and for some of the ironworkers from Batsto. In

1854 there were 22 tenant cottages, a "large store," and a tavern on the Pleasant Mills tract itself, in addition to the Elijah Clark House. Earlier the tavern had been kept by Michael Garoutte, about whom a romantic story is told. A Frenchman, he had followed Lafayette to America to help the colonial cause in the Revolution. After a skirmish in the Mullica region, Garoutte had been left for dead by his comrades. He was discovered to be alive by a girl from Pleasant Mills, who got him to her home and nursed him back to health. In story-book fashion he married her, and they settled down in a house near the old millpond. In 1825 Garoutte established a stage line to run from "Joseph English's Ferry, Camden, to Leeds Point, opposite Bearmore's Beach, which is the most convenient place at all times to pass to Long Beach, Tucker's Beach or Brigantine." Garoutte advertised that his stages left Camden at 7:00 every morning, passing through Haddonfield and Long-a-Coming and halting at Pleasant Mills for night lodging—at his own tavern, of course! These stages, the advertisement promised, reached Leeds Point "in time to pass to any of the Beaches the next day." The fare was "the same as on the Tuckerton route," which was $2. The "Tuckerton route" referred to ran over the old Tuckerton Road, through Washington, about five miles north of Pleasant Mills. These Leeds Point stages continued to run for many years.[32]

It is only possible to conjecture who was living in the Elijah Clark House in those days. In the early period of the cotton mill the occupants probably were the Lippincotts, who at first seemed to be the moving spirits of that enterprise. Benjamin Richards was busy in Philadelphia politics. Augustus Richards was a lawyer. And one did not commute to Pleasant Mills at that time. A single clue has come down to us: scratched with a diamond on the glass of a first-floor window of the old mansion are the name "William Lippincott" and the date "September 8, 1823."

Phoenix-like, a new industry soon rose from the ashes of the old Pleasant Mills cotton factory. At that time mass production of paper was still in its infancy, and demands for paper exceeded the supply of raw materials. Linen and cotton fiber, the base for hand-made paper, always had been in comparatively short supply. Now with machinery available for paper-making, all sorts of raw materials were being tried out. These included straw, bark, hemp, cat-tail stalks, ground wood, and potatoes. Several paper mills in those years were even importing shiploads of mummies from Egypt, so as to make paper from their cloth and papyrus wrappings. Pleasant Mills, however, had yet another basic source of paper close at hand: salt grass from the Jersey coastal marshes. As previously noted, this material had been used successfully for some years at McCartyville, about eight miles away, and new owners there were expanding the plant and renaming the town around it "Harris-ville." It is said, although no evidence is available, that before William McCarty established that paper mill he had enjoyed some connection with the Pleasant Mills cotton factory.

On February 8, 1860, the Pleasant Mills tract finally passed out of the Richards family. The land, the mansion, the waterpower, and the factory ruins were sold by Lewis Henry Richards to four partners: Robert M. Pierce, Benjamin F. Holbrook, John Mc-Neil, and Thomas Irving.[33] A year or so later these new owners erected a paper mill on the site of the cotton factory. From two to three dozen workers were employed, and things began to look up in the old village. The paper mill, however, seems to have en-countered financial trouble almost at the start. At first the firm name appeared as "Pierce and Holbrook," but those gentlemen were eliminated from the picture by way of a sheriff's attachment, and "Irving and McNeil" carried on. On August 7, 1862, any remaining interest of "Pierce and Holbrook" was formally con-veyed to John W. Farrell,[34] and for many years after the name "Farrell" was synonymous with prosperity for the paper mill and the community of Pleasant Mills. William E. Farrell bought his father's half interest in the paper mill on February 9, 1865, and four months later—on June 20—bought out Irving and McNeil as well, thus acquiring complete control of the now-flourishing property. This deed to Farrell, incidentally, included the grantor's interest in the "old Batsto Forge stream and ponds" and in "one hundred acres of land adjoining . . . known as the Batsto Forge Tract."

William E. Farrell, in association with his father, operated the paper plant as the "Nescochague Manufacturing Company." For some years he had a partner, Amor Hollingsworth, to whom he sold a one-third interest in 1870. After Hollingsworth's death, his heirs conveyed this share back to Farrell, in 1879.[35]

Some interesting sidelights on the operation of this first paper mill are afforded by an old "Day Book" [36] which covers the late 1860's and early 1870's. The labor force appears to have varied— in the mill proper—from 16 to 21 employees, several of whom were women. In addition to the main works there was a "bag room" which does not seem to have been in continuous operation. Only women were employed there, sometimes as many as twelve, with wages ranging from $12 to $27 monthly, probably for either part-time or piece work.

Three of the male workers—Josiah Smith, William Reed, and Henry Holloway—were employed steadily from 1867 to 1875. In those days workmen were "docked" for misconduct or carelessness. Josiah Smith, for instance, was penalized $4 for "waste of bleach and disobedience." John R. Ogden forfeited $3 for neglect of duty by shutting down the mill for high water "when there was no necessity." It cost Abe Reed $2 to violate a rule against smoking in the cutting room. In November, 1872, no less than four men were fined for "leaving the mill without notice," and Charles Adams paid $13 for "damages by breaking buggy."

As for the mill itself, the records show that in June, 1868, it was necessary to buy a "circular wheel, shaft and saw frame"; in August, 1869, two "cedar tubs" cost $20; and in April, 1871, a new boiler was hauled from Elwood station on the Pennsylvania Railroad at a cost of $44.75. The company evidently was responsible for operating the local "District School No. 43." Charles Durwan taught there from September 22, 1873, to July 4, 1874, for a salary of $380. A gross of crayons in those days cost 20 cents, and two fourth-grade readers were $1.80.

Other community projects appear on the factory records. In December, 1872, $171.63 went for "lumber and labor on flume bridge on Pleasant Mills and Batsto road." Similar work at the floodgate bridge on the road to Elwood cost $219.61. Four and one-half days' work on the dwelling house cost only $4.75. Also shown are dealings with "Atsion Mill" and Joseph Wharton.

The town itself was now prosperous and contented. Across the lake from the famous mansion were the Post Office, a general store, and a tavern, at that period kept by Samuel Kemble and his wife.

This is said to have been called the Temperance Tavern but how
Kemble induced his customers to practice temperance during the
usually merry stagecoach stopovers history does not reveal. As for
the mansion, Farrell was then the occupant of its historic rooms,
even though at that time the deed-writing lawyers described him
as a "singleman."

William E. Farrell was born in St. Louis on March 9, 1838. At
first he worked in a dry goods shop. Later he became a salesman,
which he remained until with his father he took over at Pleasant
Mills. His Nescochague Manufacturing Company had a Philadelphia office during the latter 1870's and early 1880's, first at
506–10 Minor Street and later at 609 Chestnut Street.[37] Still later
Farrell seems to have been one of the partners of Baugh, Farrell &
Warren, paper merchants in the Quaker City.

Disaster again struck Pleasant Mills in October, 1878. The paper
factory was leveled by fire. Once the flames had gained headway
the town's still-primitive fire-fighting equipment could do no more
than prevent their spreading to the houses nearby and the mansion
across the road. Fortunately the mill was insured, and Farrell
swiftly decided to rebuild. The actual reconstruction, however,
was far from swift. An old record book notes that after an insurance survey of the ruins there were "delays in completing the
mill." The following year there is a rueful comment: "Mill complete but no machinery." Even after the machinery had been
installed there seems to have been trouble. An entry for January 11,
1881, observes: "Mill not in condition to make paper," and adds:
"Savey there to correct his machinery." At last, however, operations were under way. The records contribute the following facts:
Capital was $100,000. Capacity was 4,000 pounds per day—"New
rolls produce double the old rolls." There are notes, too, concerning purchase of parts for the mill. These included dandy rolls,
press rolls, pasters, felts, stuff, and pump pulleys.

Along with the reconstruction Farrell decided to reorganize his
company. The old Nescochague Manufacturing Company now became the Pleasant Mills Paper Company, and the property was
conveyed by deed to the new concern. Farrell himself was president, and Herman Hoopes secretary. While the property changed
hands a number of times during the years that followed, hence-
forth the changes were effected through transfer of stock control

of the Pleasant Mills Paper Company and not, at least for seventy years, through transfer of title by deed to the property itself.

The owners of the new paper mill made the plant as fireproof as possible. Among other measures a lofty water tank was put up on the lawn of the mansion house, just across from the mill. In addition, two fireplugs were installed at appropriate locations along the road, while hoses and other emergency equipment were kept handy. These and other fire-prevention measures are indicated on a 1901 insurance layout map of the mill. This map also locates the various departments and shows what a large-scale undertaking the paper factory was at the height of its prosperity. The rebuilding had been a major operation; even the sluiceway from Lake Nescochague had been moved from its old location north of the mansion house to its present location on the south side. Soon the mill was rated as one of the most important industries in Atlantic County.

Unfortunately William Farrell became seriously ill at that peak of Pleasant Mills prosperity. In 1892 he made a recovery. While brief, it was of sufficient duration to be celebrated by his marriage, at the age of 54. His bride was Miss Celia Hyslop, of Troy, New York. The life together of Celia Hyslop and William Farrell proved short indeed. On March 9, 1893—his fifty-fifth birthday— Farrell died. When it was disclosed that he had left the entire Pleasant Mills estate to his new wife, trouble arose. By the terms of a previous will, others would have inherited the property, and Farrell's later will was contested in the courts. Litigation dragged on for five years. While it was in process the mill was operated by a receiver, Howard Cooper,[38] and the Elijah Clark House was occupied by William Oliver, who had been Farrell's manager and who carried on under the receiver. A new complication arose when Mrs. Farrell, before the will contest was settled, decided to remarry, her new husband being Lewis M. Cresse, a banker, and later Mayor of Ocean City.

Finally, the courts acted. They ruled in favor of the widow, who now was Mrs. Cresse. That brought about some quick decisions. Manager Oliver was replaced by Alexander J. McKeone. Mrs. Cresse made her husband president of the Pleasant Mills Paper Company, and he took over top direction from its Philadelphia office. Before long, strife and controversy were forgotten, and things went smoothly once more at the old factory.

Fire, however, remained a constant threat. At the turn of the century, it hit the village hard, although this time the paper mill was spared. In the last week of April, 1900, a forest fire started eight miles away, near Atsion. It got out of control and soon burned a two-mile swath through the forests to Batsto and Pleasant Mills. The *Philadelphia Bulletin* of May 2, 1900, reported:

> Residents of Batsto and Pleasant Mills fought the flames for two days and a night. They finally got them under control . . . but not before they had destroyed six houses on Water street in Batsto, and the Jervis homestead on the road to Washington.
>
> While the houses in Batsto were burning an alarm of fire was sounded from the Pleasant Mills Paper Mill. Boys with matches had set fire to a hay stack at the rear of Postmaster John Reynolds' home. The employes of the paper mill were all away fighting flames on the Batsto side of the river, with the exception of Manager McKeone, his son, and Fred Jervis, who was attending the engines.
>
> These men, with the assistance of an insurance inspector, ran the fire hose from the mill to the scene of the blaze but it had gained such headway that Reynolds' barn and outbuildings and those of his brother-in-law Robert Lye, were destroyed. The cinders from the burning buildings set fire to the brush land nearby . . . and now the flames flew down the west side of the river destroying everything before them.
>
> The Catholic Church, which stood in the center of a pretty little graveyard, was the first to go. . . .

Numerous other buildings were consumed by this second fire before the flames were conquered. Among these were the homes of Francis Mathis, Francis Love, and John Wilson, the American Mechanics Hall, and quite a few barns. Some of the buildings were insured, but not the church or the Jervis homestead.

When Manager McKeone took over at Pleasant Mills in 1901 he launched a badly needed improvement program. The firm's books show such entries as these: "Renewing electric wires in bleach and engine room; new cylinder in stone cutter." On August 10 repairs were made to the water wheel. These included new gears and a new shaft. That same summer "new front steps for the mansion house" were built. Later came new flooring in the "sizing room," a new "pressure regulator for the Boiling Tubs," and a new tank for the steam boiler. Incidentally, that same year the *Industrial Directory of New Jersey* noted: "PLEASANT MILLS: Population 90. Land for factory purposes may be purchased at low prices. The Pleasant Mills Paper Company has its plant here and employs about 25 persons."

The guiding spirit of Pleasant Mills in those first years of the new century was Alexander J. McKeone. With his family, he lived in the old mansion house. There from one direction he could hear the special music of his mill, and in another could gaze reflectively over the peaceful waters of Lake Nescochague. Handsome, and a gentleman of the old school, McKeone was kindly but firm as an executive, considerate of his employees and a counselor to them as well. It was McKeone who broke the tradition of running the mill all day and every day, including Sunday. By his decision the machinery was halted at midnight Saturday and not set in motion again until after midnight Sunday. In the era of the five-day week such a change may not seem impressive. For those times it was drastic. A devout Catholic, McKeone, like many members of all faiths, believed in keeping the Sabbath holy. He also sought better housing conditions for his workmen, and encouraged them to save money and open bank or building-and-loan accounts.

William Hofstetter, McKeone's son-in-law and a distinguished artist, has described to the author some of his still-vivid impressions of the busy days at Pleasant Mills. The hum of the machinery could be heard night and day, and so attuned did the ears become to it that any major deviation from normal in that "industrial symphony" was a trouble signal that brought all hands running to see what might be wrong. Paper-making was—and still is—a complicated process. Any break in the paper as it was threaded through the line of manufacture could result in havoc, if the machinery were not halted promptly. In those days the mill was making quantities of kraft or wrapping paper, as well as documentary paper for the Federal Government. Another product was a special grade of stock used as a base for sandpaper and emery paper. These wares were hauled to the Elwood station of the Pennsylvania Railroad, five miles away; and the teams would bring back goods for the mill and the village which had arrived at the Elwood freight platform. The tempo of living in Pleasant Mills was even and slow, the citizens' wants and pleasures were simple, and the mill—

economic heart of the town—enjoyed a steady prosperity. Here was a happy community by any standard.

Then came 1914, and three things happened. World War I broke out; President Lewis M. Cresse died suddenly; and Mrs. Cresse decided to close down the paper mill. No single-industry town could withstand such blows, and Pleasant Mills soon felt their full impact. The formal shutdown of the paper plant took place in 1915, after 34 years of continuous operation. Closing day was a dark one for the village generally, but a particular tragedy for the two dozen families whose men had worked at the mill and now found their livelihood gone.

Two years later Manager McKeone acquired control of the Pleasant Mills Paper Company and ownership of its 300-odd acres of land. Despite the war and wartime shortages and restrictions, he made a valiant attempt to reopen the mill. For raw material he tried rope hemp. He even revived the use of salt hay from the Mullica marshes. It was useless. Coal and transportation were but two of the tough industrial problems born of the war; and with a location so far from the rail lines it became difficult for McKeone even to assure delivery of such paper as he was able to manufacture. Soon the plant closed again.

Previously, on September 5, 1915, as the owner of the Pleasant Mills Paper Company, McKeone—by deed—put the old mansion house and the lake site around it in his own name.[39] Thus the Elijah Clark House—or "Kate Aylesford House"—was separated from its famous mill site for the first time in a century and a half. It was a fortunate decision on McKeone's part. Had he not made it, there is a good chance that the mansion would have been neglected and let go to ruin, a fate which soon befell the paper mill. After this transfer of the dwelling, McKeone sold stock control of the mill property to the Norristown Magnesia & Asbestos Company. The new proprietors had ambitious plans. Installing special machinery, they set out to produce asbestos. While the plant ran for several years, this experiment did not prove profitable. Later an attempt was made to resume manufacture of paper, especially colored paper for theater tickets, with newspaper waste as the raw material. This venture also failed. Finally, in April, 1925, the Norristown firm closed the paper mill. They removed most of the machinery, and junkmen carried off the metal scrap. The old factory now was not only idle; it was empty. Its days as an industrial enterprise were over.

In the later 1920's efforts were made to develop the Pleasant Mills tract as a residential club community. These languished. Long years passed. Decay set in. Bit by bit the old mill crumbled. The roof of the main building collapsed. Floors sagged and timbers rotted. Soon much of the structure was wide open to the assaults of the elements, and within twenty years a modern paper mill had come to look for all the world like a century-old ruin. By 1945 the mill was a wreck, and the whole property was put up for tax sale by the Township of Mullica. On July 31, 1945, it was purchased by the McCorkle-Pleasant Mills Company,[40] which is still promoting it as a residential real estate development.

Alexander McKeone deeded the Elijah Clark House to two of his daughters, Katherine E. McKeone and Elizabeth V. Rafferty, on December 7, 1925. On January 25, 1928, McKeone passed away, at 77. Thus closed another chapter in the Pleasant Mills story. It was an end to one more era. A new generation, with new ideas, was coming up, and art would take over—or endeavor to—where industry had left off.

A first step in this new direction was sale of the southern portion of the mansion site to Raymond and Mary Baker in 1946. Mary Baker was the daughter of Frances McKeone and William Hofstetter. The Bakers were both professional artists, and, with Mr. Hofstetter, they built a charming single-story home close by Lake Nescochague. Their aunts were living in the old mansion itself, and right across the road were the ruins of the old mill, at which the Bakers often cast interested eyes. They perceived its artistic possibilities. Hopes and plans took shape. Finally, on October 21, 1948, they bought the shell of the old mill and the ground adjacent. In the deed from the McCorkle-Pleasant Mills Company, there is this unusual restriction:

> The buildings now erected upon the said land . . . shall be preserved and kept in their present artistic appearance and such repairs shall be made thereto as shall make said buildings usable for the purpose hereinafter mentioned. . . .

> The said lands and premises shall be used only as an art center for the promotion and teaching of art, crafts and the manufacture

and sale of art craft products . . . provided that the tenant or
tenants shall have the right to live and make their home in said
buildings.[41]

This purpose was to be realized in a surprising way by Mrs. Ada
Fenno, of Swarthmore, Pennsylvania, a widow with philanthropic
interests and a fondness for the theater. She was fascinated by the
romantic ruins of the old mill, and by the improvements which
the Bakers were making. Her wish was to develop the place as a
playhouse. That wish became reality when the Bakers sold her the
property on January 18, 1952.[42] More money and more work were
needed before the place was fully playworthy, but all that was
necessary was accomplished, and the "Mill Playhouse," as it now
was named, opened its doors on July 3, 1953. Its first offering consisted of three plays from Noel Coward's *Tonight at Eight-Thirty*:
"Family Album," "Ways and Means," and "Red Peppers."

The Pleasant Mills story closes on a particularly pleasant note:
the purchase of the Elijah Clark House from their aunts by Raymond and Mary Baker. Transfer of the house took place on
November 17, 1954. The Bakers will keep the venerable mansion
they love in the family to which it has meant so much for so long.
To paraphrase a verse of the late James Lane Pennypacker:

> In the spirit of those pioneers
> Rededicate
> This house to inspiration for the years
> And to a kindly fate.

THE HUNDRED YEARS

An even century has passed since closing of the Batsto cupola marked an end to the making of iron in the pines of New Jersey, and sent the founders, molders, and forgemen far away to newer smelters with richer ore-fields. Today, in one of history's curious turnabouts, the iron industry is returning in force to the Delaware River Valley. One giant enterprise, the United States Steel Corporation's "Fairless Works," is in large-scale operation at Morrisville. The National Steel Corporation has acquired a 2,600-acre tract near Paulsboro as the site for a new steel mill. Late in 1956 the Phoenix Iron & Steel Company purchased a thousand acres above Burlington for construction of a "fully integrated plant." Others may follow. And these huge modern mills will lie only a few moments by motor car from the sites of their rustic predecessors—the bog-iron furnaces, with their little thirty-foot stone stacks.

Those hundred years have presented tremendous challenges to human ingenuity and imagination. Henry Bessemer invented his converter for making steel cheaply in the very year that Batsto's last smelter fires were dying out. Men had been fabricating steel by crude processes and in limited quantities since before the time of Christ, but not until Bessemer's was there any process capable of large-scale production. That was only a start. Ever-increasing demands of the machine age brought a succession of new steel-making techniques, notably the open hearth process. Still, the production pace never seemed quite rapid enough. The American steel industry which sprang from such small beginnings to world preeminence celebrates its centenary in 1957, but economic pressure for more and more iron products has never ceased accelerating since Charles Read built his little forges and furnaces in the woods nearly two hundred years ago.

When Read started out, the New Jersey bog ores seemed abundant. Despite their reputed capacity for natural reproduction over a twenty-year cycle, most deposits were exhausted in less than fifty years. Even before that, the Jersey furnaces had begun obtaining ores from other states whose resources at the time seemed plentiful. So it has gone through the hundred years. Those Pennsylvania furnaces whose competition doomed their Jersey counterparts soon found their own production demands soaring beyond the supplies of nearby mines. They, too, began to import. As recently as 1892 the great Mesabi range was opened in Minnesota. Here, some thought, was a truly inexhaustible source of the vital mineral; yet with close to two billion tons already extracted, depletion of the Mesabi deposits seems near at hand.

Thus our nation, which once boasted of self-sufficiency in iron—of all metals—today is importing iron ore in increasing quantity from abroad. As the scows used to bring both ore and pig iron up the Mullica River to the hungry furnace at Batsto, great ore-boats now move in never-ending procession up the Delaware River to the great new steel plants of the valley. Those boats bear ore from faraway Venezuela; and meanwhile the steel mills of Western Pennsylvania are being fed with ore hauled all the way from Labrador.

Is there still iron in the pines? Can ore be found today in the lakes, streams, and bogs? It can. Another twist of history, and men may be mining it again—for purposes we cannot foresee, on a tomorrow which may be closer than we think.

THE BATSTO STORE BOOKS
(Selected Extracts)

1851

JANUARY
3 THR [Thomas H. Richards] left for Mt. Holly & Phila.
7 THR came home.
13 Mr. Richards [Jesse] very unwell.
15 Schooner "Frelinghuysen" arrived with corn.
21 Shipping glass.
23 Shipping glass and boat boards.
27 THR started for Medford.

FEBRUARY
7 Chalkley Leek burnt his hand.

MARCH
1 Teams at shider woods.
7 Snow storm. Teams all idle.
11 Town meeting at Quaker Bridge.
19 Saw mill stopped by back water.

APRIL
14 Shipping glass.
15 Finished loading the "Frelinghuysen."
21 Show of wild beasts at P. Mills. ½ day of hands lost.

MAY
29 Roofing grist mill.
30 Fire in glass works this day.

JULY
   15 Mr. R. went to Phila stage.
   29 Fire in Sleepy Creek swamp.

AUGUST
   5 Hauling logs and moulding sand.
   14 Made 1st melt in Cupola.
   21 Cupola idle.
   25 Mr. & Mrs. Richards went to the beach.

SEPTEMBER
   8 Commenced blowing glass in both houses.
   14 Cupola idle for want of coal.

OCTOBER
   Cupola worked all month.

NOVEMBER
   4 Election at Quaker Bridge.
   7 Repairing casting house.
   [Cupola worked 16 days in November.]

DECEMBER
   5 Salting pork, burning lime, cleaning out casting house.
   17 Carting boards from Hammonton.
   21–24 Snow.

1852

JANUARY
   Cupola worked 10 days.

FEBRUARY
   Cupola worked 13 days.

MARCH
   9 Town meeting at Quaker Bridge.
   [Cupola worked 19 days in March.]

MAY
   6 Building new Cupola.
   7 Taking down clay mill & lath mill.

8 Taking down bellows.
11 Cleaning out for new Cupola.
21 Building Cupola.
22 Finished bridge at flume.
29 Building floor to grist mill.

SEPTEMBER
9 Building new house on Bridge street.
24 Idle. Hands to camp meeting.
28 Mr. R. to Tuckerton. Cattle show.

NOVEMBER
8 Election at A. Nichols [Crowleytown tavern].

DECEMBER
17 Cupola idle for want of Scotch pig iron.
27 Surveyors running road to R.R. Taking in pork.

1854

MAY
13 Carting glass to "Mary" [vessel].

JUNE
17 Mr. Richards demise 8 o'clock a.m.
18 All the work on the premises stopped from 16th to 21st on account of the death of the proprietor.

AUGUST
3 Sold lot of cedar swamp near Hammonton.

SEPTEMBER
15 Surveying at P. Mills.
16 Husking corn. Finished loading the "J. Wurts."
21 Cupola finished Scotch iron.
22 Mrs. Daniel Reid killed by fall from wagon at store.

OCTOBER
5 Repairing P. Mills church. Surveying at Quaker Bridge.

NOVEMBER
14 Cupola at work. Grist mill broken.

DECEMBER
16 Repairing furnace wheel. Cupola worked 4 days. Want of pig
   iron.

1855

JANUARY
19 Fixing safe. [Cupola idle all month. No pig iron.]

FEBRUARY
19 "Ida" and "Mary" from N.Y. [Cupola idle all month.]

APRIL
4 "Ida" from N.Y. Pig iron and goods.
  [Cupola worked 3 days in April.]

MAY
26 "Ida" and "Mary" arrived in ballast. [No ore.]
   [Cupola worked 7 days in May. No pig iron.]

JUNE
[Cupola did not work all month.]
Store robbed of cash about $66 on the night of 13th or morning
of 14th. [Cash included 167 three-cent pieces.] Returned by
A.S. Doughty [who worked in the store!].

JULY
[Cupola idle. Glass house idle. Five men working.]

AUGUST
8 Fire near Hammonton. Sent men to it. Fixing lath mill.
10 Fire near Wilson houses. Men at it till 2 P.M.
11 Men partly idle from fatigue fighting fires. Cupola idle. Glass
   house idle.

OCTOBER
12 Show at P. Mills. Hands partly idle.
25 Tombstone from Mt. Holly.
26 Monument erected over Mr. R's grave.
27 3 men loading the "Laura" with wood.
28 Fire at Penn Swamp.

NOVEMBER
6 Election at Green Bank.
15 The "Ida" from N.Y. Lime stone.
27 Cotton factory at P. Mills burnt down.

1856

JANUARY
1 Meeting of church trustees. THR here.
5 Snow 16 inches deep.
8–12 Mills all frozen up.
22–25 Mills all frozen up. C.B. Henderson superintending glass works.

FEBRUARY
6 Put 40 hams in pickle.
13 J. Campion came with beef.
18 Teams at work. Mills idle.

MARCH
7 Lath mill idle for want of belts.
10 Circular [saw] mill idle.
11 Town meeting at Green Bank.
12 Mills all at work.
[Cupola idle. One glass house full-time.]

JUNE
3 Carting scrap iron from P. Mills to "Mary."
13 Fire out in glass house.

JULY
8 Saw mill broken down.
11 Surveying lines between Batsto and Atsion.
[Cupola idle. Glasshouse idle.]

NOVEMBER
6 Fitting belt on saw mill.
13 Saw mill running.
21 Carting hay.
24 Finished carting wood. Pulling turnips.
25 SPR and his lady arrived here.
[The glasshouse had been idle all year from June 13, the cupola idle since May 11, 1855.]

1857

JANUARY
2 Mills at work. The "J. Wurts" in, light.
6 Wm. R. Braddock surveying near Washington Tavern.

JUNE
2 Started lath mill.
22 Furnace stone came by "Rebecca."

JULY
14 Commenced building glass furnace.
27 Put cap on furnace.

AUGUST
25 Put fire in glass furnace today.

SEPTEMBER
5 Collision on RR 8 p.m. near W.H. Station. T. & SPR hurt.
8 Blowing glass.
10 Flattening oven out of order.
25 Shipping glass to the "Wurts."
26 Shipping glass and boards to "Ida."
28 THR & SPR to Phila.
[October to December: Glasshouse full-time.]

1858

MARCH
31 Carting lumber to Hammonton.

APRIL
1 Carting slabs to Hammonton. [Now using R.R.]
8 Carting glass to Hammonton.
14 Shipping glass to the "Rebecca."

JUNE
4 Glass factory blew out. Sloppy glass.
5 Finished shipping glass. Shipping boards.
17 Carting cedar logs. Ploughing. SPR home.

JULY
   3 Heavy rain and thunder P.M.
  12 Sodding grave yard.
  19 Ploughing corn.

AUGUST
  23 Put fire in glass house furnace 6 P.M.

SEPTEMBEᴋ
   7 Two teams to Medford. Repairing flood gate bridges. Commenced blowing glass.

1859

JANUARY
13–15 Glass house idle for want of dry wood.

FEBRUARY
  Glass house full time.

MARCH
  10 Sawing in both mills.

JUNE
   8 Ploughing corn and potatoes.

JULY
  Glass house idle all month.

AUGUST
   1 Carting scrap iron to landing.
   2 Taking down Cupola. THR gone west.
   3 Tearing down Cupola.
  18 Repairing roof of factory.
  24 Commenced blowing glass.
  28 Aurora Borealis apparent.

NOVEMBER
   8 Election at Lower Bank.

1860

FEBRUARY
  1 Stage stopped on account of snow.
    [Glasshouse full-time January to April.]

MAY
  17 Surveyor laying out race to P. Mills. Five hands at cranberry
     patch.

JUNE
  25 Taking down glass furnace.

JULY
  20 Building furnace. Glass house idle all month.

AUGUST
  27 Commenced blowing glass.

SEPTEMBER
  5 J. Fitzgerald died.

OCTOBER
  2 Sowing rye. Cutting hay.
  24 Sawing in both mills.
  29 Glass to the "Catherine."
  30 Sale of timber near Washington.

NOVEMBER
  21 Planting apple trees.

DECEMBER
  9 Snow all day.
    [Glasshouse full-time.]

1861

JANUARY
  14 Filling ice house.

FEBRUARY
  13 Carting glass to Hammonton station.

MAY
10 Planting potatoes, cutting shiders.
24 Hay from Harrisville.

JUNE
29 Fire in glass house.

SEPTEMBER
19 Made 1st blowing of glass.
26 No outside work being Fast Day, by President's Proclamation.

NOVEMBER
13 Shipping glass on boat "Rebecca."

DECEMBER
[Glasshouse full-time.] Business seems to be falling off.

1862

MAY
31 Glass house lost 1 blowing on account of Home Guards going to Tuckerton.

JUNE
20 Tearing down old Glass Furnace.

JULY
4 Teams all idle. No work in the place.

AUGUST
25 SPR to Washington.
26 Fire in flattening house.

DECEMBER
19 Glass to the "J.B. Cramer."

1863

JANUARY
5 Circular saw at work.
30 Sawing in both mills.
[Glasshouse full-time to the 24th.]

APRIL
  24 Heavy rain. Great Freshet.
  30 Thanksgiving Day.

JUNE
  1 Fire in woods.
  2 Hands fighting fire in woods.
  3 Fire in woods.
  17 Tearing down oven.
  18 Tearing down furnace.
  [Glasshouse worked 12 days. Thirteenth fire in the glasshouse.]

JULY
  2 Taking down old house.
  20 Labor trouble.

AUGUST
  31 Blowing glass this day.

SEPTEMBER
  24 Cranberries to station.
  25 Glass to Atsion.

NOVEMBER
  [Carting glass 14 days to Atsion, to Jersey Southern R.R.]

DECEMBER
  [Glasshouse full-time.]

APPENDIX II

STREET DIRECTORY OF BATSTO
1853–1856

BRIDGE STREET

R. Moore
Ben Ford

C. B. Henderson
Joshua Cline
Wm. R. Lutis
T. Henderson

Wm. T. Williams
D. McAneny
C. Leek
B. Lloyd
Saml. Smith
D. Neippling
F. Frolinger
L. Hirsch
Joseph Scull
Richard Scull
John Hart
J. Murphy
Eph. Scull
J. Lloyd
Joab Hugh
N. Kell
R. Kell
H. Birdsall
S. Brown
N. Neippling
A. Pricket
A. Hankins
J. S. Barret

OAK STREET

Benj. Camp
Geo. Reeves
F. Smith
R. Stewart
S. Smallwood
J. Mickel

WATER STREET

S. McAneny
T. Baxter
S. Jervis
Jos. Patterson
Jas. McAneny

M. Moore
Widow Lipsett
R. M. McLoughlin
John Stickel
H. Kroup
J. McAneny
Reuben White
Daniel Brown
Joseph Bird
Wm. Patterson
Chas. Brewer
Samuel Scull
Wm. Alexander
Thos. Dunlap
Jas. Patterson
John Shields

TUCKAHOE STREET

C. Taylor
C. Lamphey
Wm. Anderson
Jno. Ford
N. Shi . . . . . .
J. Froelinger
R. Wall

CANAL STREET

John Peterson
H. Stewart
C. Camp
D. Camp
H. D. Rinehart
J. Fitzgerald
G. Reinhart
Geo. M . . . . . .
D. Southard
John Cox

APPENDIX III

RICHARD WESTCOTT * TO CHARLES READ

This indenture made the third day of May 1765 Between Richard
Westcott of Egg Harbour, yeoman and Margaret his wife on the one
part and Charles Read of Burlington on the other part. Witnesseth
that the said Richard Westcott and Margaret his wife for and in Con-
sideration of the Sum of Two hundred pounds to the said Richard
Westcott in hand paid by the said Charles Read the receipt of which
is hereby acknowledged, They the said Richard Westcott and Margaret
Have Granted and Sold and by these presents do Grant, bargain and
sell unto the said Charles Read, his heirs and Assigns One undivided
half part of a Saw Mill with two saws wch the said Richard Westcott
bought and was conveyed unto him by Daniel Ellis, Sheriff of Burlington County and by Robert Friend Price Sheriff of the County of
Gloucester as the Estate which a certain John Fort had conveyed to
him by John Munrow and Vincent Leeds by Indenture of the Eighth
of August 1761 and which Indenture from the said Sheriffs bears date
the ninth day of May 1764. Together with all the Lands, Houses
Streams Damms and Appurtenances in the afsd Indentures Expressly
excepting a Certain piece of Cedar Swamp known by the name of John
Forts or Joseph Tashesold Works and which is sold to Elias Gandy
and does not go higher up the Swamp than Lovelands Northwest
Corner and also a piece of pine land lying in the fork or point of
Batstow and Atsion [rivers] and is [located] by the streams of sd rivers
and a Strait Line Drawn from Batstow at the Indian field about
thirty rods above a house where Jesse Freeman lives about a quarter of a
mile from the meeting of the Two Rivers and Extending to a point
onto Atsion at a Bend of the said river just across the Blacksmith's shop
and about a quarter of a mile from the Grantor's house wherein he
now lives and the Timber on which Land and the right of Fishing and
Erecting Fishing Sheds on the shore of Batstow is intended to be
hereby Granted to the said Charles Read his heirs and assigns. Together with all the privileges and Appurtenances to the granted premises belonging. TO HAVE AND TO HOLD the above premises and
Tracts of Land and premises to the said Charles Read his heirs and
Assigns to the only proper use and behoof of the said Charles Read

* So spelled here.

his heirs and assigns forever. In testimony whereof the parties here-
unto set their hands and Seals the Day and Year first above written.
Richard Westcott [seal] Margaret Westcott [seal] Sealed and Delivered. . . .

Endorsed and Be it Remembered that on the third day of May
Anno Domino One Thousand Seven Hundred and Sixty-five Richard
Westcott and Margaret his wife the Grantors in the within Indenture
mentioned being to me well known personally appeared before me
John Ladd, Esq., One of His Majesties Council for the province of
New Jersey and severally acknowledged that they Sealed and Delivered
the within Indenture as their Several Acts and Deeds to the uses and
purposes therein mentioned and the said Margaret being by me
privately Examined apart from her Husband did acknowledge that She
Sealed and Delivered the Same freely and voluntarily without any
fear threat or compulsion. John Ladd.

"The Doggs" Chains . . . and implements to the mile belonging
or therewith used go with the premises within granted as freely as if
they passed by Express words in the indenture. Richd Westcott.

Recorded May 31, 1761

APPENDIX IV

AN ACT
For the Preservation of Cranberries

Passed the 10th of November 1789

Whereas it has been represented to the legislature that cranberries,
if suffered to remain on the vines until sufficiently ripened, would be a
valuable article of exportation;

Therefore

Be it enacted by the Council and General Assembly of this State,
and it is hereby enacted by the authority of the same, That if any person or persons shall, after the passing of this act, take or gather from
the vines, at any time after the first day of June, and before the tenth
day of October, cranberries on the common or unlocated lands within
this state, or on any lands not their own property, or for which they

pay no tax, such person or persons shall forfeit and pay, for every such offence, the sum of twenty shillings, and also, the further sum of twenty shillings, for every bushel so taken or gathered within the times aforesaid, and so in proportion for a greater or lesser quantity, to be sued for and recovered in any court where the same may be cognizable, with costs of suit, to be applied, one half, if on the common or unlocated lands, to and for the use of the county where the offence shall have been committed, or if on any of the located lands, one half to be paid to the owner or possessor of said land, and the other half, together with the cranberries, so as aforesaid taken and gathered, to the use of the persons who shall sue for and recover the same.

CHAPTER REFERENCES

ABBREVIATIONS

AC: Atlantic County deeds, housed in Mays Landing
BC: Burlington County deeds, at Mount Holly
GC: Gloucester County deeds, at Woodbury
NJ: Deeds and records in the Secretary of State's Office in Trenton
SGO: Deeds or records, in the Surveyor General's Office, of the Council of West Jersey Proprietors, Burlington, N.J.
WE: Deeds or records in the possession of the Wharton Estate

CHAPTER 1
1. John W. Harshberger, *The Vegetation of the New Jersey Pine Barrens* (Philadelphia, 1916), p. 2.
2. *Ibid.*, p. 4.
3. Camden *Post Telegram*, November 24, 1915.
4. Kenneth Fisk, "A Tale of Lands and Buildings," *Review of the Society of Residential Appraisers* (Chicago), Vol. XXII, No. 2 (February, 1956), pp. 16–22.

CHAPTER 2
1. E. N. Hartley, *Hammersmith, 1643–1675: The Saugus Iron Works Restoration* (Saugus, Mass., 1955), p. 26.
2. Batsto Store Books.
3. Charles S. Boyer, *Early Forges and Furnaces in New Jersey* (Philadelphia, 1931), p. 4.
4. *Ibid.*, p. 13.
5. BC deeds Liber C, p. 12.

6. Boyer, *Early Forges and Furnaces in New Jersey,* p. 45.
7. WE records.
8. Dennis C. Kurjack, *Hopewell Village National Historic Site* (Washington, D.C., 1952), p. 9.
9. John F. Hall, *The Daily Union History of Atlantic City and County* (Atlantic City, 1900), p. 472.

CHAPTER 3
1. Carl Raymond Woodward, *Ploughs and Politicks: Charles Read of New Jersey and His Notes on Agriculture* (New Brunswick, 1941), p. 24.
2. *Ibid.,* pp. 89–90.
3. NJ Liber AF, p. 179.
4. *Ibid.*

CHAPTER 4
1. SGO deeds Liber S 6, p. 293.
2. NJ deeds Liber AC, p. 180.
3. Charles S. Boyer, *Early Forges and Furnaces in New Jersey* (Philadelphia, 1931), p. 167.
4. K. Braddock-Rogers, "The Bog Ore Industry in South Jersey Prior to 1845," *Journal of Chemical Education,* Vol. VII, No. 7 (July, 1930), p. 1496.
5. GC deeds Liber B, p. 128.
6. Letter of Micajah Mathis to Rep. George Sykes, Jan. 20, 1847.
7. BC deeds.
8. John F. Watson, *Annals of Philadelphia and Pennsylvania in the Olden Time* (Philadelphia, 1898), Vol. II, p. 543.
9. Camden *Post,* January 18, 1894.
10. Boyer, *Early Forges and Furnaces in New Jersey,* pp. 172–173.
11. Henry Charlton Beck, *Forgotten Towns of Southern New Jersey* (New York, 1936), p. 275.
12. John F. Hall, *The Daily Union History of Atlantic City and County* (Atlantic City, 1900), p. 31.
13. BC Liber Q 5, p. 472.
14. BC Liber R 6, p. 447.
15. BC Liber V 6, p. 23.
16. Wheaton J. Lane, *From Indian Trail to Iron Horse* (Princeton, 1939), p. 355.
17. Camden *Democrat,* May 8, 1875.
18. E. M. Woodward and John F. Hageman, *History of Burlington and Mercer Counties* (Philadelphia, 1883), p. 451.
19. Camden County *Courier,* June 23, 1888.

CHAPTER 5
1. George R. Prowell, *The History of Camden County N.J.* (Philadelphia, 1886), p. 353.
2. Alfred M. Heston, *Absegami* (Camden, 1904).

3. BC deeds Liber X 7, p. 52.
4. BC wills Book C, p. 308.

CHAPTER 6
1. K. Braddock-Rogers, "Fragments of Early Industries in South Jersey," *Journal of Chemical Education*, Vol. VIII, Nos. 10, 11 (Oct., Nov., 1931), p. 1926.
2. BC Liber F, p. 202.
3. Hopewell Furnace records.
4. BC Liber D 3, p. 623.
5. Braddock-Rogers, "Fragments of Early Industries in South Jersey," p. 1930.
6. *Ibid.*, pp. 1926–1928.
7. BC deeds Liber G 6, p. 26.
8. *Ibid.*
9. WE records.
10. BC Book of Incorporation, Liber C, p. 18.
11. BC Liber X 11, p. 516.
12. WE records.
13. *Ibid.*
14. BC Liber 324, p. 99.

CHAPTER 7
1. BC Liber C, p. 494.
2. K. Braddock-Rogers, "The Bog Ore Industry in South Jersey," *Journal of Chemical Education*, Vol. VII, No. 7 (July, 1930), p. 1506.
3. Charles S. Boyer, *Early Forges and Furnaces in New Jersey* (Philadelphia, 1931), p. 115.
4. BC deeds Liber T, p. 676.
5. Batsto Furnace books.

CHAPTER 9
1. NJ deeds Liber Z, p. 186.
2. SGO deeds Liber W, p. 474.
3. NJ deeds Liber U, p. 284.
4. NJ deeds Liber U, p. 289.
5. NJ deeds Liber X, p. 265.
6. NJ Sessions Laws, 1766.
7. Charles S. Boyer, *Early Forges and Furnaces in New Jersey* (Philadelphia, 1931), p. 176.
8. *Pennsylvania Packet*, April 15, 1784.
9. NJ deeds.
10. Boyer, *Early Forges and Furnaces in New Jersey*, p. 167.
11. NJ deeds, Liber AC, p. 189.
12. Boyer, *Early Forges and Furnaces in New Jersey*, pp. 181–183.
13. *Ibid.*, p. 183.
14. *Ibid.*, p. 178.
15. WE deeds.

16. *Ibid.*
17. Marquis James, *Biography of a Business, 1792–1942: Insurance Company of North America* (Indianapolis, 1942), pp. 17–18.
18. Atlantic County Historical Society records.
19. Richard P. McCormick, *Experiment in Independence: New Jersey in the Critical Period, 1781–1789* (New Brunswick, 1950), p. 129.
20. James, *Biography of a Business*, p. 57.
21. WE records.
22. Louis Richards, *Sketches of Some Descendants of Owen Richards* (Philadelphia, 1882), p. 11.
23. BC deeds Liber L, p. 336.
24. BC deeds Liber C, p. 336.
25. George De Cou, *The Historic Rancocas* (Moorestown, N.J., 1949), p. 134.
26. BC wills Book C, p. 342.
27. Dennis C. Kurjack, *Hopewell Village National Historic Site* (Washington, D.C., 1952), p. 28.
28. Batsto Store Books.
29. *Ibid.*
30. Bertram Lippincott, *An Historical Sketch of Batsto, New Jersey* (n.p., 1933), p. 11.

CHAPTER 10

1. George De Cou, *The Historic Rancocas* (Moorestown, N.J., 1949), p. 199.
2. NJ deeds Liber AF, p. 179.
3. Charles S. Boyer, *Early Forges and Furnaces in New Jersey* (Philadelphia, 1931), p. 166.
4. *Ibid.*, p. 157.
5. Pemberton Papers.
6. Cadwallader Papers.
7. Carl Raymond Woodward, *Ploughs and Politicks: Charles Read of New Jersey and His Notes on Agriculture* (New Brunswick, 1941), pp. 407–411.
8. BC deeds Liber T, p. 314.
9. BC deeds Liber T, p. 317.
10. Ballinger family records.
11. BC deeds Liber F 6, p. 88.
12. Boyer, *Early Forges and Furnaces in New Jersey*, pp. 166–167.
13. *The Medford Lakes Story* (Medford Lakes, N. J., 1952), p. 6.

CHAPTER 11

1. Carl Van Doren, *Secret History of the American Revolution* (New York, 1941), pp. 172–175.
2. *Ibid.*, pp. 244–246.
3. Alfred M. Heston, ed., *South Jersey: A History, 1664–1924* (New York, 1924), Vol. II, pp. 759–761.

4. Charles F. Green, *Pleasant Mills and Lake Nescochague—A Place of Olden Days* (Hammonton, N. J., n.d.), p. 8.
5. *Ibid.*, p. 10.
6. SGO Liber M, p. 268.
7. Atlantic County Historical Society records: Biographies and Family Records of Revolutionary Soldiers.
8. *Ibid.*
9. SGO Liber S 6, p. 137.
10. Charles S. Boyer, *Stage Routes in West New Jersey* (Camden, 1935), p. 6.
11. Atlantic County Historical Society records.
12. Heston, *South Jersey: A History*, Vol. I, p. 226.
13. *Ibid.*
14. Atlantic County Historical Society records.
15. *Ibid.*
16. *Ibid.*
17. Green, *Pleasant Mills and Lake Nescochague*, p. 15.
18. Heston, *South Jersey: A History*, Vol. II, pp. 768–769.
19. GC deeds Liber TT, p. 396.
20. Marquis James, *Biography of a Business, 1792–1942: Insurance Company of North America* (Indianapolis, 1942), pp. 69–70.
21. GC deeds Liber M, p. 248.
22. GC deeds Liber QQ, p. 399.
23. Green, *Pleasant Mills and Lake Nescochague*, p. 17.
24. Clevenger Papers.
25. *Ibid.*
26. WE records.
27. NJ Laws 1824, p. 62.
28. GC deeds Liber TT, p. 494.
29. Trenton *Federalist*, October 18, 1824.
30. GC deeds Liber TT, p. 496.
31. GC mortgages Book M, p. 62.
32. Boyer, *Stage Routes in West New Jersey*, p. 8.
33. AC deeds Liber M, p. 488.
34. AC deeds Liber Y, p. 524.
35. AC deeds Liber AC 72, p. 441.
36. Nescochague Manufacturing Company records.
37. *Ibid.*
38. Green, *Pleasant Mills and Lake Nescochague*, p. 22.
39. AC deeds Liber 587, p. 43.
40. AC deeds Liber 1235, p. 496.
41. AC deeds Liber 1414, p. 205.
42. AC deeds Liber 1596, p. 433.

BIBLIOGRAPHY

MANUSCRIPTS

The following deeds and records have been consulted. The abbreviations given in parentheses are those used in the chapter references.

Atlantic County deeds, housed in Mays Landing (AC); Burlington County deeds, at Mount Holly (BC); Gloucester County deeds, at Woodbury (GC); deeds and records in the Secretary of State's Office in Trenton (NJ); deeds and records in the Surveyor General's Office of the Council of West Jersey Proprietors, in Burlington (SGO); deeds or records in the possession of the Wharton Estate, and since turned over to the State of New Jersey (WE).

Other manuscript sources consulted are noted below:

Atsion Furnace Books: Burlington County Historical Society.
Batsto Store Books: Privately owned.
Clark Family Notes: Atlantic County Historical Society.
Clevenger Papers: Atlantic County Historical Society.
Elting Papers: Historical Society of Pennsylvania.
Frambes, Harriet, Old Stage Line Records: Atlantic County Historical Society.
The Aaron Leaming Diary: Historical Society of Pennsylvania.
The Martha Furnace Diary: Copies of Nathaniel R. Ewan and Captain Charles Wilson.
Pemberton Papers: Historical Society of Pennsylvania.
Pennington Family notes: Atlantic County Historical Society.
Samuel Richards papers: Historical Society of Pennsylvania.
Robert Stewart papers: Owned by Mr. and Mrs. John Stewart, of Nesco.
Wescoat records: Atlantic County Historical Society.

MAPS

1683: Seller, John and Fisher, William. A Map of New Jersey in America.
1719: "I. Senex." A New Map of Virginia, Maryland, and the Improved Parts
    of Pennsylvania and New Jersey.
1749: Evans, Lewis E. A Map of Pensilvania, New-Jersey, New-York and the
    Three Delaware Counties.
1777: Faden, William. The Province of New Jersey.
1778: Romans, B. A Chorographical Map of the Country Round Philadelphia.
1800: Carey. American Pocket Atlas. New Jersey.
1828: Gordon, Thomas. Map of New Jersey.
1831: Finley, A. Map of New Jersey.
1834: Gordon, Thomas. Map of New Jersey: History of New Jersey.
1839: Rogers, Henry D. New Jersey Geological Map.
1849: Otley, J. W. and Whiteford, R. Map of Burlington County, N. J.
1856: Saunders, E. H. Map of New Jersey.
1858: Kuhn and Janney: New Map of Burlington County.
1860: Stone, C. K. Map of New Jersey.
1860: Map of Oswego Tract, formerly Martha Survey.
1872: Beers, Comstock and Cline. New Jersey Atlas.
1876: Scott, J. D. Combination Atlas of Burlington County, N. J.
1877: Hopkins, G. M. Map of New Jersey.

PRINTED SOURCES

Barber, John W., and Howe, Henry, *Historical Collections of the State of New
    Jersey*. New York, S. Tuttle, 1844.
Beck, Henry Charlton, *Forgotten Towns of Southern New Jersey*. New York,
    E. P. Dutton & Company, 1936.
Blackman, Leah, "Historical Sketch of Tuckerton." *Proceedings of the Sur-
    veyors Association of West New Jersey* (1880).
Boyer, Charles S., *Early Forges and Furnaces in New Jersey*. Philadelphia,
    University of Pennsylvania Press, 1931.
——, *Indian Trails and Early Paths*. Camden, N.J., Camden County
    Historical Society, 1938.
——, *Stage Routes in West New Jersey*. Camden, N.J., Camden County
    Historical Society, 1935.

Braddock-Rogers, K., "Fragments of Early Industries in South Jersey." *Journal of Chemical Education*, Vol. VIII, Nos. 10, 11 (Oct., Nov., 1931), pp. 1915–1929.

———, "The Bog Ore Industry in South Jersey Prior to 1845." *Journal of Chemical Education*, Vol. VII, No. 7 (July, 1930), pp. 1493–1519.

Cawley, James S. and Margaret, *Exploring the Little Rivers of New Jersey*. Princeton, Princeton University Press, 1942.

Cazenove, Theophile, *Cazenove Journal, 1794*, ed. by Rayner Wickersham Kelsey. Haverford, Pa., The Pennsylvania History Press, 1922.

Craig, C. Chester, *Council of Proprietors of West Jersey*. Camden, N.J., Camden County Historical Society, 1922.

De Cou, George, *Moorestown and Her Neighbors*. Philadelphia, Harris and Partridge, 1929.

———, *The Historic Rancocas*. Moorestown, N.J., printed by the *News Chronicle*, 1949.

Drinker, Elizabeth, *Extracts from the Journal of Elizabeth Drinker: From 1759 to 1807*, ed. by Henry D. Biddle. Philadelphia, J.B. Lippincott Company, 1889.

Dunbar, Seymour, *A History of Travel in America*. Indianapolis, Bobbs-Merrill Company, 1915.

Eberlein, Harold Donaldson, and Hubbard, Cortlandt Van Dyke, *Portrait of a Colonial City*. Philadelphia, J.B. Lippincott Company, 1939.

Gordon, Thomas F., *A Gazetteer of the State of New Jersey*. Trenton, D. Fenton, 1834.

[Green, Charles F.], *A History of Pleasant Mills Church*. n.p., n.d.

Green, Charles F., *Pleasant Mills and Lake Nescochague—A Place of Olden Days*, 3d ed. Hammonton, N.J., privately printed, n.d.

Hall, John F., *The Daily Union History of Atlantic City and County*. Atlantic City, N.J., The Daily Union Printing Company, 1900.

Harshberger, John W., *The Vegetation of the New Jersey Pine Barrens*. Philadelphia, Christopher Sower Company, 1916.

Hartley, E.N., *Hammersmith, 1643–1675: The Saugus Iron Works Restoration*. Saugus, Mass., published by the Restoration, 1955.

Heston, Alfred M., *Absegami: Annals of Eyren Haven and Atlantic City, 1609 to 1904*. Camden, N.J., privately printed, 1904. 2 vols.

———, ed., *South Jersey: A History, 1664–1924*. New York, Lewis Publishing Company, 1924. 4 vols.

Hillman, Sarah Crawford, *Historical Sketch of Potter Street in Haddonfield*. Haddonfield, N.J., Haddon Gazette Press, 1910.

James, Marquis, *Biography of a Business, 1792–1942. Insurance Company of North America*. Indianapolis, Bobbs-Merrill Company, 1942.

Kobbé, Gustav, *The New Jersey Coast and Pines*. Short Hills, N.J., G. Kobbé, 1889.

Kurjack, Dennis C., *Hopewell Village National Historic Site*. National Park Service Historical Handbook Series, No. 8. Washington, D.C., 1952.

Lane, Wheaton J., *From Indian Trail to Iron Horse*. Princeton, Princeton University Press, 1939

Lee, Francis Bazley, ed., *Genealogical and Memorial History of New Jersey.* New York, Lewis Historical Publishing Company, 1910.

Lineage Book of the National Society of Daughters of Founders and Patriots of America. Washington, D.C., 1910–.

Lippincott, Bertram, *An Historical Sketch of Batsto, New Jersey.* n.p., privately printed, 1933.

Lippincott, William R., *Traditions of Old Evesham Township.* Moorestown, N. J., reprinted from the Moorestown *Republican*, 1911.

McCormick, Richard P., *Experiment in Independence: New Jersey in the Critical Period, 1781–1789.* New Brunswick, Rutgers University Press, 1950.

Medford Lakes Story, The (25th Anniversary pamphlet). Medford Lakes, N.J., 1952.

Moore, Harvey, *An Old Jersey Furnace.* Baltimore, Newth-Morris Printing Company, 1943.

New Jersey Archives. Newark, New Jersey Historical Society, 1880–.

Peattie, Donald Culross, *A Natural History of Trees of Eastern and Central North America.* Boston, Houghton Mifflin Company, 1950.

Pennypacker, James Lane, *Verse and Prose.* Haddonfield, N.J., The Historical Society of Haddonfield, 1936.

Peterson, Charles J. *Kate Aylesford: A Story of the Refugees.* Philadelphia, T.B. Peterson, 1855.

Prowell, George R., *The History of Camden County, N.J.*, Philadelphia, L.J. Richards & Company, 1886.

Richards, Louis, *Sketches of Some Descendants of Owen Richards.* Philadelphia, Collins, Printer, 1882.

Spargo, John, *Iron Mining and Smelting in Bennington, Vermont, 1786–1842.* Bennington, Bennington Historical Museum, 1938.

Stewart, Frank H., ed., *Notes on Old Gloucester County, N.J.*, Camden, printed by Sinnickson Chew and Sons, 1917.

Stone, Witmer, *The Plants of Southern New Jersey With Special Reference to the Flora of the Pine Barrens: Report of New Jersey State Museum.* Trenton, 1910.

Van Doren, Carl, *Secret History of the American Revolution.* New York, The Viking Press, 1941.

Watson, John F., *Annals of Philadelphia and Pennsylvania in the Olden Time.* Philadelphia, E.S. Stuart, 1898. 2 vols.

Werner, Charles J., *Eric Mullica and His Descendants: A Swedish Pioneer in New Jersey.* New Gretna, N.J., privately printed, 1930.

Woodward, Carl Raymond, *Ploughs and Politicks: Charles Read of New Jersey and His Notes on Agriculture.* New Brunswick, Rutgers University Press, 1941.

Woodward, E.M., and Hageman, John F., *History of Burlington and Mercer Counties, N.J.* Philadelphia, Everts and Peck, 1883.

W.P.A., *New Jersey, A Guide to Its Present and Past.* American Guide Series, New York, The Viking Press, 1939.

INDEX